Food Hygiene, Health and Safety

J Audrey Stretch
BSc (Hons), C. Biol, M.I. Biol

H A Southgate
Cert Ed.

Addison Wesley Longman Limited
Edinburgh Gate, Harlow,
Essex CM20 2JE, England

First published in Great Britain 1991
Second impression 1994
Third impression published by Addison Wesley Longman Limited 1997
Reprinted 1998

British Library Cataloguing in Publication Data
Stretch, J. Audrey
Food hygiene, health and safety.
I. Food. Hygiene
I. Title II. Southgate, H.A.
363.1927

ISBN 0–273–03386–7

10 9 8 7 6 5 4 3 2

Produced through Longman Malaysia,PP

Contents

Preface

This text has been designed to accompany practical instruction in Food Hygiene and Safety in colleges and industrial training courses. It provides the necessary background knowledge of Food Hygiene and Safety legislation, food related illnesses and safety procedures. The book covers the hygiene syllabus requirements of the IEHO, RSPH&H, RSH and BTEC Hotel and Catering operations examinations.

The central theme is the importance of establishing safe routines for the storage, preparation, cooking and service of food to protect both caterer and client.

Students are encouraged to analyse the problems associated with food handling through hazard analysis control methods and their grasp of hygiene is tested in the food poisoning case histories in the last chapter.

Acknowledgements

The authors and publishers would like to thank the following for providing and giving permission to reproduce illustrations in this book.

Electronic Temperature Instruments Ltd
Killgerm Ltd
The National Dairy Council
The Science Photo Library
Rentokil
W & G Sissons Ltd

Cover photograph reproduced by kind permission of Dr Tony Brain; Science Photo Library.

1 Introduction to hygiene, health and safety

The catering industry employs more than 1.25 million people in around 300 000 outlets around the UK and is a complex and diverse industry. On the profit side employees work in hotels, restaurants, pubs, clubs, fast food outlets and on British Rail and airlines. On the service side, caterers work in office and factory canteens, prisons, hospitals, the school meal service and meals on wheels for the elderly. They provide a variety of cuisines - traditional British, French, Italian, Indian, Chinese, Mexican and many more - to satisfy the tastes of an increasingly multiculturally aware and multi-ethnic population.

Catering is a growth industry which between 1981 and 1986 showed an annual addition to the workforce of about 20 000. It offers full and part time employment to people of all ages and levels of skill, from the highly skilled chefs in first class hotels and restaurants to young school leavers in sandwich and hamburger bars. It is an industry which responds to changes in taste and life styles, so that as one type of catering declines, another rises to take its place. Although a very traditional industry, there is also an increasing use of technology in the central production of meals for airlines, hospitals and schools, through cook-chill, cook-freeze and sous-vide processes.

Around one in eight meals in this country are now consumed outside the home, so the catering industry shoulders a heavy burden for the nutritional health and safety of the nation.

Basic legislation relating to catering

Because food is a basic necessity for everyone, the State recognises that legislation is needed to control the standards of food offered for sale and for the preparation and service of food. The most important pieces of legislation which affect the catering industry are:

- The Food Safety Act 1990
- The Food Hygiene (Amended) Regulations 1989
- The Health and Safety at Work Act 1979

The Food Safety Act 1990

The Food Safety Act amalgamates much of the separate statutes relating to food into one Act covering all parts of Great Britain. The intention of

this statute is to ensure safety throughout the food chain, from its source in farms, market gardens and fisheries until it reaches the consumer.
The Food Safety Act seeks to protect the consumer from potential danger, using a variety of powers of control:

- It protects against being offered harmful or unfit food.
- It protects against being offered substandard food – 'not of the nature, substance or quality' expected by the purchaser.
- It safeguards against the false or misleading labelling of food.
- The Act regulates the composition and the standards of hygiene expected of food supplied to the public.
- It makes the *possession* and not just the sale of unfit food, an offence.
- It requires all food premises to be registered with the Local Authority. Registration will allow notices to be served quickly to remedy any defects in food premises or, where there is thought to be a serious risk to the public, to close down the business without delay.
- The Act has powers to control contaminants which might get into food, e.g. traces of animal medicines in meat.
- It allows Ministers to make emergency orders to deal with potentially serious problems relating to food, e.g. the import of adulterated wine or cooking oil.
- It makes provision for food handlers to be trained in food hygiene. Local Authorities may provide hygiene courses and may charge for this service.
- The Act also takes powers to control *new* methods of production and processing of foods, e.g. the use of irradiation to control bacterial growth and insect infestation of foods.

The Food Hygiene (Amended) Regulations 1989

The Food Hygiene Regulations cover the practical details of food premises, food preparation and food handling personnel. The main areas covered are:

- Preparation, cooking and storage of foods, both raw and cooked.
- Personal hygiene of all food handlers.
- Maintenance of all rooms used for food preparation and storage.
- Requirements and conditions for staff changing and rest rooms.
- Requirements for washing hands, equipment and crockery.

Environmental Health Officers (EHOs) are empowered to enforce the provisions of the Food Safety Act and the Food Hygiene Regulations.

The Health and Safety at Work Act

This Act seeks to protect both employers and employees in the working environment. The main areas of concern are:

- The maintenance of premises and machinery.
- Standards of lighting, safety notices and fire evacuation notices.
- Employees' duties and responsibilities to the employer.

Health and Safety Officers have the power to enforce this Act.

The necessity for legislative control of food businesses

Hygiene

The necessity of strict regulations controlling food production is brought home to everyone when outbreaks of food illness are reported in the media. They may concern 'new' diseases such as Listeriosis which rarely affected people in the past but increased in incidence when there were changes in the type of food eaten or the methods of preparation or storage. Alternatively, the increase in food illness may be due to a well-known food poisoning organism changing its habits, as in the Salmonella-in-eggs problem when the organism began infecting the egg laying apparatus of hens.

These sudden changes in food hazards remind us that we can never afford to relax standards of food hygiene. In fact, the statistics of reported cases of bacterial food poisoning cannot give anyone who works in the food industry any cause for complacency, because the numbers affected have risen steadily in recent years.

Table 1.1 Laboratory reported cases of bacterial food poisoning in England and Wales, 1986–8

	Year		
	1986	*1987*	*1988*
Salmonellae (all types)	14 177	16 991	23 100
C. Perfringens	896	1 266	1 312
S. Aureus	67	178	111

Not all the food illnesses reported in Table 1.1 were contracted from catering premises, many were reported from family outbreaks, associated with poor hygiene and storage in the home, but, of the general outbreaks, the greater number were from receptions, restaurants and hotels. Obviously, there is a need for strict legislation and training in food hygiene for *all* food handlers whether full or part time, permanent or temporary.

Safety

Catering employees, at all levels, encounter hazards in the course of their work since it involves handling raw food which may be infected, the use of knives, slicing and cutting mach. ery, cooking equipment, and lifting and stacking stores. Any of these activities can cause ill health or injury to the unwary or inexperienced worker.

Table 1.2 Place of general food poisoning outbreaks reported by laboratories, MOsEH & EHOs in England and Wales, 1986–8

Place of outbreak	*Salmonella sp*	*Clostridium perfringens*	*Staphylococcus aureus*	*Bacillus cereus & bacillus sp*	*Total*
Restaurant/ reception/hotel	158	55	7	33	253
Hospital	41	33	1	5	80
Institution	26	31	3	1	61
School	13	4	-	-	17
Community	18	-	-	-	18
Shop	18	-	7	3	28
Canteen	9	15	1	2	27
Farms	5	-	-	-	5
Infected abroad	25	1	-	-	26
Other	15	5	1	1	22
Unspecified	40	8	6	1	55
Total	368	152	26	46	592

How safe is the hotel and catering industry

During the year 1988–89 280 non-fatal major injuries and 1747 injuries requiring more than 3 days' absence from work were reported to the relevant authorities. These figures do not put the catering industry in the high risk category with mining, oil extraction and the construction industries, but they do represent an unnecessary amount of suffering and loss of working hours which could be reduced by improved working conditions and higher levels of training. The Health and Safety at Work Act is designed to establish working conditions which reduce the occurence of accidents and safeguard the health of employees.

For any kitchen team to be efficient in the preparation and production of meals, whether in a small domestic kitchen or a large commercial concern, safe methods and practices must be learnt and used. Not only is the *learning* of skills important, but also the ability to *adapt* them to the various working situations in the kitchen. The development of awareness of the correct procedures of good workmanship are of prime importance to the chef, cook and their assistants. The Food Safety Act 1990 and the Food Hygiene Regulations (Amended) 1989 state that food offered for sale must be free from contamination and fit for human consumption. So food must be prepared hygienically in clean surroundings, and the food handlers themselves must maintain a high standard of personal hygiene.

Fig 1.1 Areas of danger in the kitchen

Implementing the food legislation

If the kitchen staff are to prepare food to an acceptable standard, the kitchen must be planned to enable them to work efficiently. The layout must provide continuity of work flow from the point the raw materials are received, to the point of serving or dispatching the cooked food. The work flow must ensure staff movement is kept to a minimum to promote efficiency and to reduce the cause of accidents. Kitchens must be planned and designed with hygiene in mind. The materials used for the walls, ceilings, floors and work surfaces (*see* Table 1.3) and the design of equipment should allow for thorough and easy cleaning (*see* Fig 1.2(a)). There must be sufficient, suitably placed, facilities for hand and machine washing of equipment and crockery; wash hand basins for the staff and available storage areas for all equipment.

Kitchen planning

The basic principles of kitchen layout, as far as safety and hygiene are concerned, include the following points:

- The need for adequate ventilation.
- The requirement for good natural lighting, if possible. Where artificial light has to be employed, strip lighting is preferable, with diffusers (shades).
- The type and variety of heavy equipment needed must be considered, with preference being given to the design and siting which will allow for thorough and easy cleaning.
- The layout must ensure work flow of raw materials from receiving point to the service of the cooked food.
- Ample space is needed for storage of dry and perishable goods.
- The layout must facilitate easy and quick removal of swill and refuse to the appropriate short term storage areas.

Food Hygiene Regulations

22 Suitable and sufficient means of lighting shall be provided in every food room and every such room shall be suitably and sufficiently lighted.

23 Except in the case of a room in which the humidity or temperature is controlled, suitable and sufficient means of ventilation shall be provided in every food room and suitable and sufficient ventilation shall be maintained there.

Fig 1.2(a)

Fig 1.2(b)

Task

Compare the two pictures of kitchens in Fig 1.2.

1 Which kitchen would you rather work in?

2 Note the differences in:
  (a) ventilation and lighting
  (b) design of equipment
  (c) standard of hygiene and cleanliness.

Figure 1.3 shows the workflow in a kitchen emphasising the hygiene regulations and showing the equipment sinks and hand and food sink requirements rather than specific cooking equipment.

Kitchen surfaces

All surfaces for walls, ceilings and working surfaces must be suitable for the purpose. The angles where the walls and ceilings and walls and floors meet, must be concave, to prevent the accumulation of dirt, dust, grease and food particles and to allow for easy and thorough cleaning. Heavy equipment must be positioned to give access all round for efficient cleaning.

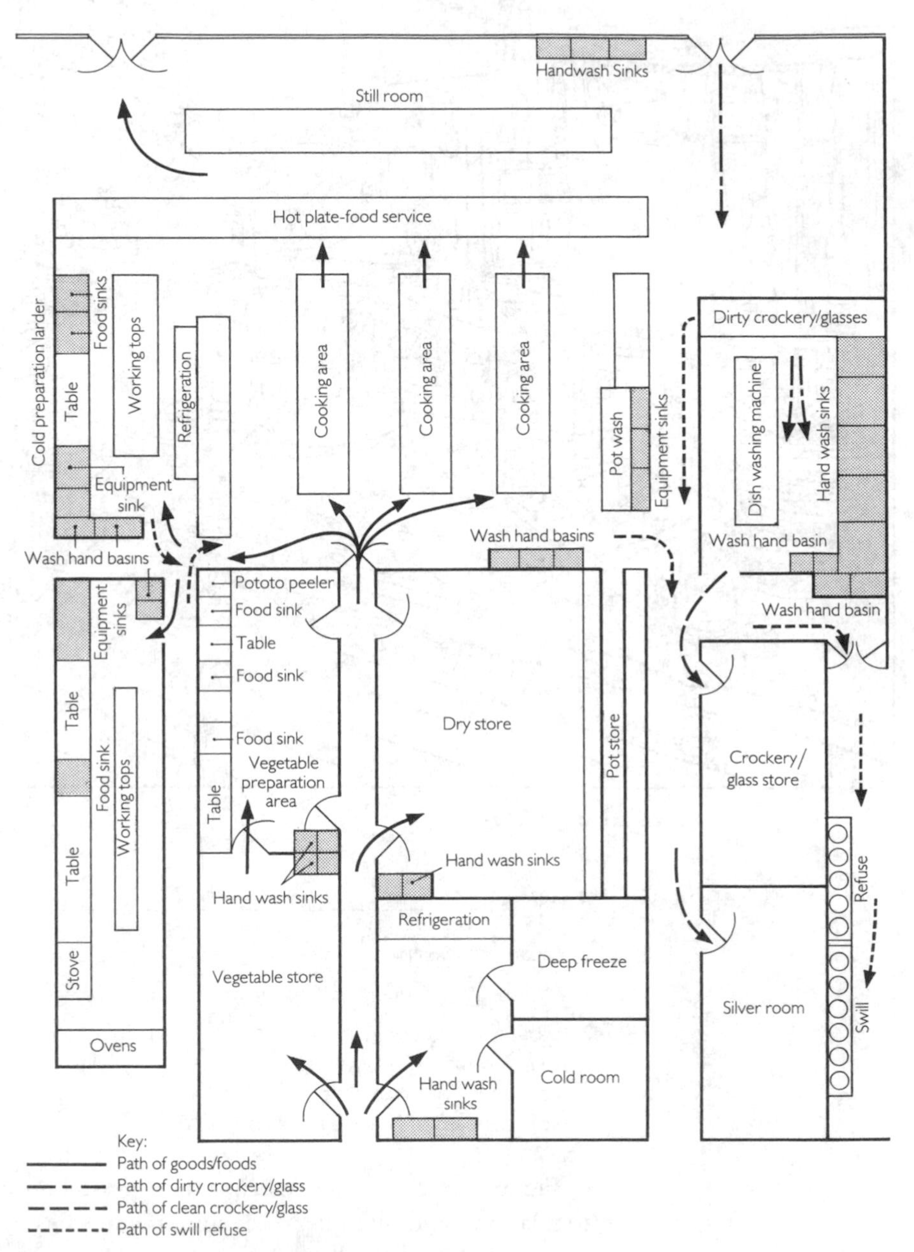

Fig 1.3 Kitchen planning diagram to show work flow

Food Hygiene Regulation

25 The walls, floors, doors, windows, ceiling, woodwork and all other parts of the structure of every food room shall be kept clean and shall be kept in such good order, repair and condition as to

(a) enable them to be effectively cleaned; and

(b) prevent, so far as reasonably practicable, the entry of birds, and any risk of infestation by rats, mice, insects or other pests.

Table 1.3 Kitchen surfaces – uses and cleaning methods

Material surface	*Cleaning method*	*Uses*
Quarry tiles	Hot detergent water scrub Rinse and dry	Kitchen floors
Terrazo	Hot detergent water scrub Rinse and dry	Kitchen, food room floors
Granolithic chips	as above	Kitchen, food room floors
White glazed tiles	as above	All food room walls
Plaster, emulsion painted	as above	All food room walls
Thermoplastic	as above	All food room walls
Stainless steel	as above, do not use abrasives	Walls and tables
Formica	as above, do not use abrasives	Table tops
Hard wood, Beech or Oak	coarse salt, scrub with wire brush, remove salt	Butcher blocks Cutting blocks
Plastic	Hot detergent water Rinse dry	Cutting boards
Compressed rubber	as above	Cutting boards Butchers blocks
Marble	as above	Pastry making

Wash hand basins

Wash hand basins must be strategically placed to encourage proper hand washing when necessary. These must be equipped with:

- liquid soap dispenser
- plastic backed nail brush
- hot and cold running water
- hot air drier or paper towels
- covered bin for hygienic disposal of paper towels
- a displayed notice: THIS BASIN MUST ONLY BE USED FOR WASHING HANDS

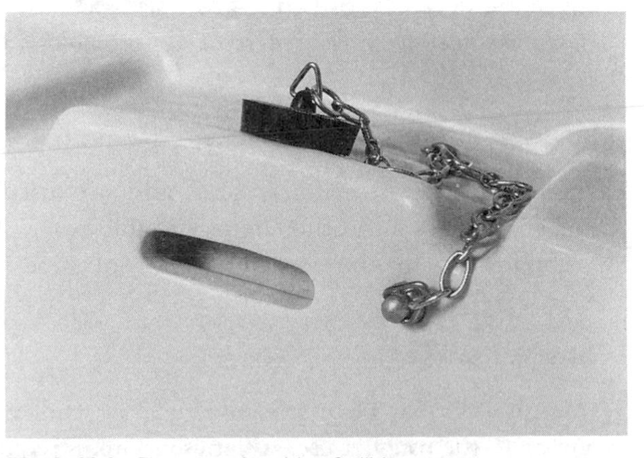

Fig 1.4(a) Clean handwashing facilities

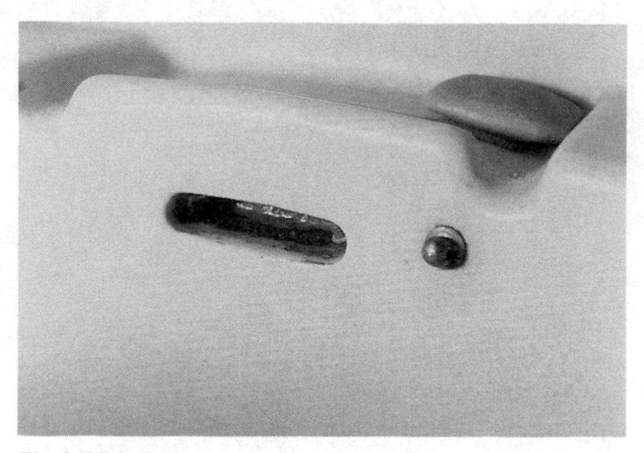

Fig 1.4(b) Dirty handwashing facilities *Courtesy: Rentokil*

Food Hygiene Regulation

18 (1) There shall be provided in all food premises suitable and sufficient washhand basins for the use of all persons engaged in the handling of food in or about those premises and such basins shall be placed in a position conveniently accessible to such persons.

(2) There shall be provided for every such washhand basin an adequate supply of hot and cold water or of hot water at a suitably controlled temperature, or in the case of food premises where no open food is handled, of cold water.

(3) There shall be provided for use at every such washhand basin an adequate supply of soap or other suitable detergent, nail-brushes and clean towels or other suitable drying facilities.

(4) Every such washhand basin shall be kept clean and in good working condition.

(5) The washing facilities provided under this regulation shall not be used for any purpose other than for securing the personal cleanliness of the user.

Food preparation sinks

Modern food preparation sinks are constructed of stainless steel. They must only be used for food preparation e.g. washing and preparing vegetables, potatoes, fish and at times, meat.
These sinks must never be used for hand washing or for washing equipment.

Hand washing equipment sinks

These sinks are of stainless steel and are used to hand wash cutlery, crockery, glasses and other items of table ware. There must be a sink, in working order, for the final sterilisation of these items.

Pot wash sinks

Pot wash sinks may be of galvanised iron or stainless steel of a thicker gauge to stand up to the rougher use. A sterilising sink is required for the final process.

Sinks used for hand washing any equipment must have a supply of hot and cold water and be in working order.

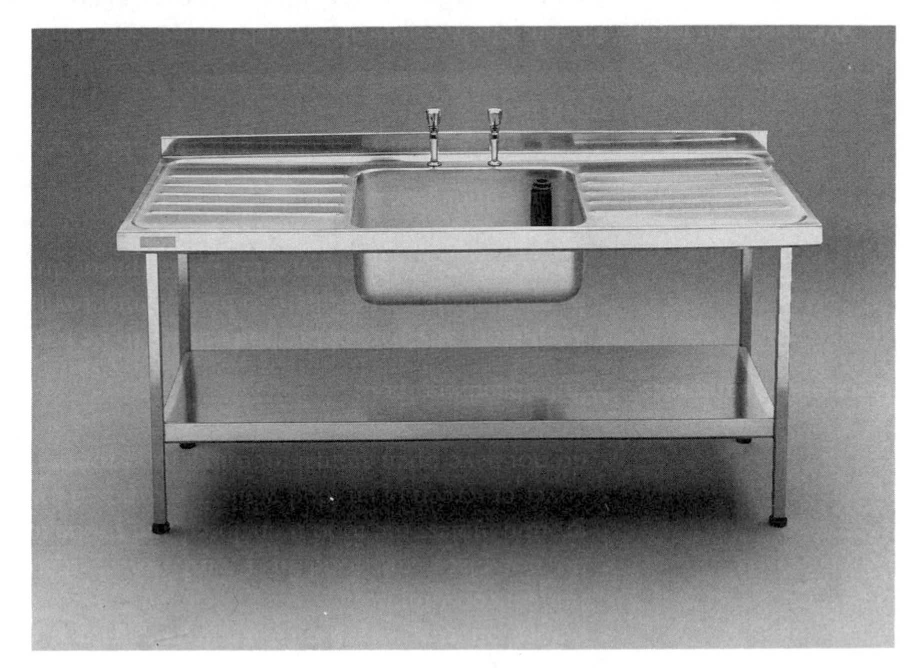

Fig 1.5 Food preparation sink Courtesy: W & G Sissons Ltd

Food Hygiene Regulation

21 (1) There shall be provided in all food premises where open food is handled sinks or other washing facilities suitable and sufficient for any necessary washing of food and equipment used in the food business; and in this regulation reference to a sink shall include a reference to any other suitable washing facility.

(2) There shall be provided for every such sink an adequate supply of hot and cold water or of hot water at a suitably controlled temperature, or cold water only where the sink is used:

(a) only for washing fish, fruit or vegetables; or

(b) for washing with a suitable bactericidal agent only drinking vessels, or only ice cream formers or servers.

(3) Every such sink shall be kept clean and in good working condition.

Staff changing rooms

There must be adequate provision for suitable staff changing rooms which are situated away from all food storage and preparation areas. All toilets must be situated away from food rooms. Each member of staff requiring a uniform i.e. chefs, cooks, assistant cooks, kitchen porters, stores clerks and still room maids must be provided with a locker or suitable area for storing their outdoor clothes and kitchen uniforms. Food must never be kept in these clothing lockers. The changing rooms must be fitted with hand basins and may have toilets and shower cubicles in an adjacent room.

All toilets must display a 'NOW WASH YOUR HANDS' notice.

The changing and toilet rooms must be well lit, have sufficient ventilation and be kept clean at all times. A rest room should be provided, preferably near the changing rooms.

Food Hygiene Regulations

16 (1) Every sanitary convenience situated in, or regularly used in connection with, any food premises

(a) shall be kept clean and in efficient order;

(b) shall be so placed that no offensive odours therefrom can penetrate into any food room.

(2) Any room or other place which contains a sanitary convenience shall be suitably and sufficiently lighted and ventilated and shall be kept clean.

(3) No room which contains a sanitary convenience shall be used as a food room.

(4) No food room which communicates directly with a room or other place which contains a sanitary convenience shall be used for the handling of open food.

(5) There shall be fixed and maintained in a prominent and suitable position near every sanitary convenience provided or made regularly available for use by persons employed in the handling of food in or about food premises, a clearly legible notice requesting users to wash their hands after using the convenience.

18 (1) There shall be provided in all food premises suitable and sufficient washhand basins for the use of all persons engaged in the handling of food in or about those premises and such basins shall be placed in a position conveniently accessible to such persons.

20 (1) Subject to the provisions of any certificate of exemption, there shall be provided in all food premises where open food is handled suitable and sufficient accommodation for outdoor or other clothing and footwear shall not be kept in any place on or about the premises other than in the accommodation so provided.

10 A person who engages in the handling of food, shall while so engaged –

(a) keep as clean as may be reasonably practicable all parts of his person which are liable to come into contact with the food;

(b) keep as clean as may be reasonably practicable all parts of his clothing or overclothing which are liable to come into contact with the food.

24 (1) No food room shall be used as a sleeping place.

(2) Subject to the provisions of any certificate of exemption, no food room which communicates directly with a sleeping place shall be used for the handling of open food.

Exercise

1 What are the requirements for:

(a) hand washing equipment sinks?

(b) wash hand basins?

(c) washing of pots and pans?

(d) sinks for food preparation?

2 What facilities should be provided by employers for the storage of outdoor clothing and personal property of food handlers?

3 Why is it necessary for all sanitary rooms to be situated away from food areas?

4 List the important points which are required for health and safety in changing rooms.

Personal hygiene for food handlers

Hand care

You work with your hands and, therefore, you must ensure that they are *always clean* with well kept nails of suitable length. You must:

- make sure your hands and nails are clean, as dirt and food particles harbour bacteria, which can be passed on to food.
- never wear nail varnish when handling food, as the varnish may flake off into food.
- keep your finger nails short, never longer than the finger tips, as long finger nails make handling food difficult.
- cover all cuts and abrasions with a *blue waterproof dressing*, while you are handling food.
- not wear jewellery when handling food. Rings with stones harbour bacteria and rings or stones may also find their way into food. Earrings which are large or of the pendant type are dangerous because they could get caught in machinery or drop into food.

Hand washing

Hands must be washed:

- after using the toilet.
- before commencing work in the kitchen.
- after handling raw food – meat, poultry, vegetables and fish.
- after handling swill or refuse.
- before preparation of cold foods which will not require further cooking.
- after smoking.

Correct method of hand washing

Hands must be washed thoroughly – not just a skin rinse.

- If warm water is supplied – wash under running water with the plug out of the basin. If hot and cold water is supplied separately – use plenty of hand hot water in the bowl.
- Washing should be slow and thorough. Take care to cover all parts of your hands and the wrists also. Rinse thoroughly in *clean* water.
- Pay particular attention to your finger nails, to remove meat, fat, fish and pastry particles which collect underneath.

Task

Cover the whole of both hands with a perfumed hand cream or hand gel with a food colour added. Wash and rinse as usual. Can you detect any perfume or food colour on any part of your hands? If so, try again.

Food Hygiene Regulation

10 A person who engages in the handling of food, shall while so engaged –
(a) keep as clean as may be reasonably practicable all parts of his person which are liable to come into contact with the food;
(c) keep any open cut or abrasion on any exposed part of his person covered with a suitable waterproof dressing.

Smoking, taking snuff

You must:

- never smoke in any area where food is stored, prepared or served (servery area). Ash may find its way into food but, what is more serious, bacteria passed from the lips to the fingers may contaminate food.
- always wash your hands after smoking during breaks before handling food.
- never take snuff in a food area or when wearing your kitchen uniform. When taking snuff the fingers come into contact with the nostrils so small particles of germ laden snuff could settle on food or work surfaces.

Food Hygiene Regulation

10 (e) [food handlers must] refrain from the use of tobacco or any other smoking mixture or snuff while they are handling any open food or are in any food room in which there is open food.

Sickness, illness

If you are sick or ill and suffering from a stomach disorder, a cough or a cold you *must*:

- inform your employer or supervisor and stop handling food.
- consult your doctor if the condition persists.
- not handle food. If you are suffering from coughs, colds or sneezes the germs are harmful to the health of others and are transferable to food.

Food Hygiene Regulation

13 (1) Immediately a person engaged in the handling of food becomes aware that he is suffering from, or is the carrier of, typhoid, paratyphoid or any other salmonella infection or amoebic or bacillary dysentery or any staphylococcal infection likely to cause food poisoning, he shall immediately notify the appropriate medical officer of health accordingly:

Provided that where the person required to give such information is himself the person carrying on the food business he shall give the information immediately to the appropriate medical officer of health.

General points of personal hygiene

- Highly scented perfume and body sprays must not be used when preparing or handling food because certain foods pick up these odours and become unpleasant to eat.
- You must *never spit*, particularly in an area where food is prepared, cooked, or served. Spitting is a filthy and dangerous habit at any time.

Food Hygiene Regulation

10 (d) [food handlers must] refrain from spitting;

9 A person who engages in the handling of food shall, while so engaged, take all such steps as may be reasonably necessary to protect the food from risk of contamination...

Exercise

1 List the common unhygienic personal habits which are not acceptable in food handling establishments.

2 Explain the importance of personal hygiene, particularly as you are a food handler.

3 What action must be taken in the event of a food handler becoming ill?

Kitchen uniform

It is important for all staff involved in food preparation to be correctly dressed. You must not wear your ordinary clothes when handling food; likewise you must not wear your kitchen dress or clothes when you are not on duty in the kitchen. There are two reasons for these rules - if you wear your ordinary clothes in the kitchen, they pick up cooking odours particularly from frying. More importantly, your ordinary clothes can bring bacteria into the kitchen and so may your kitchen uniform if you wear it when you are off duty.
The correct dress for kitchen staff consists of:

- blue and white check trousers
- white double breasted chef's jacket
- neckerchief
- chef's hat
- white food handler's apron

The modern trend is for female staff to wear the same uniform as male staff with either the chef's hat or mob-cap.

The items of uniform should be made of 100 per cent cotton because cotton is absorbent, hardwearing, comfortable to wear and can stand frequent laundering at high temperatures.

The trousers should be loose fitting around the ankles and legs as this prevents unnecessary injury to the legs if hot liquids are spilt. It also allows for the trousers to be removed easily without adding to the injury. The double breasted jacket gives extra protection to the chest from splashing liquids and from stoves and ovens. The neckerchief absorbs perspiration and it should be changed when necessary – maybe twice daily.

The chef's hat should fit comfortably on the head, enclosing the hair. This prevents the hair from absorbing cooking odours and stops hair or dandruff falling into the food. Disposable hats are available but these need changing frequently as they easily become soiled.

The food aprons should be worn below the knees to give protection to the legs. All items of the uniform should be changed when they become soiled, either through food stains and splashes or perspiration. It is sensible to wear a cotton vest when working in a hot kitchen while insulating the body when you are working in cooler preparation areas.

The kitchen cloth or rubber completes the uniform and this again should be cotton. It is used for handling hot dishes and saucepans. It must not be used for drying your hands as this makes it wet and so reduces its insulating properties. Handling hot dishes with wet cloths may cause severe burns. This cloth must not be used to handle food or shaping omelettes. Would you like to eat an omelette which had come in contact with a kitchen cloth?

Kitchen footwear must be suitable to cover the feet; with stout uppers, reinforced toe caps and non-slip soles. Stout uppers give protection from spillage and the non-slip soles give a good grip on the kitchen floors which are often slippery. The soles should be of materials which give reasonable wear against grease and the detergents used for washing the floors. The toe-caps give protection from heavy items which may be dropped e.g. chopping boards, butcher's cleavers, knives and heavy saucepan. It is clear that sandals, training shoes, plimsolls, court shoes, etc. are unsuitable for kitchen use. The shoes you choose must fit comfortably because you will be on your feet for up to eight hours on a normal working day and sometimes even longer. It is advisable to 'break in' your new shoes before wearing them in the kitchen.

Exercise

1 **(a)** What is the most suitable material for making items of kitchen uniform?
(b) Why is it the most suitable material?

2 What are the requirements of your kitchen footwear?

3 List the items of your kitchen uniform when you are correctly dressed for work?

4 State the purpose of each item of clothing.

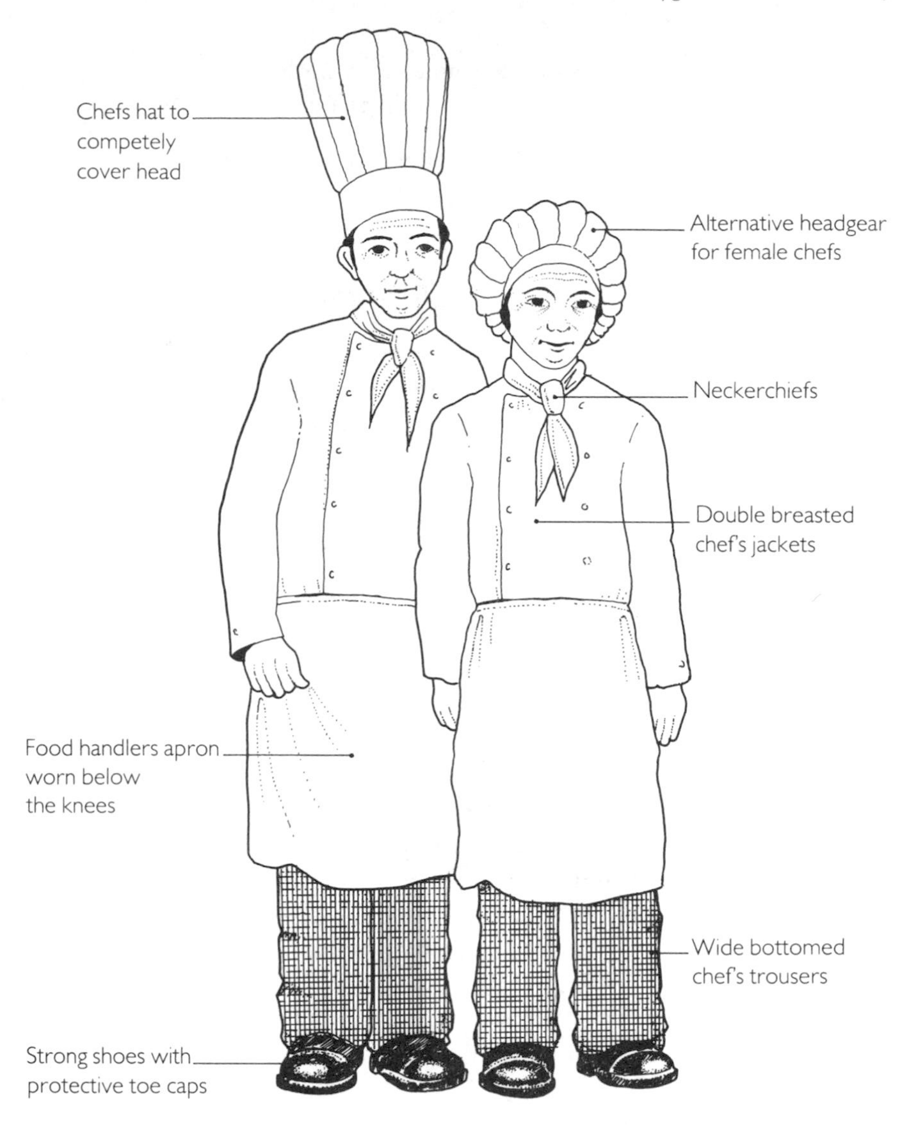

Fig 1.6 Male and female kitchen staff uniform

Legislation affecting food production premises Food Hygiene (Amended) Regulations 1990

Frequent reference has been made throughout this chapter to these important regulations drawing your attention to specific areas of the legislation. The following is a *summary* of how the provisions relate to the day-to-day running of food premises.

Part I Definitions

Food is anything sold for human consumption including drinks, chewing gum, sweets, etc., but not:

- milk
- water, live animals or birds
- any substance sold as a medicinal drug, e.g. Aspirins, Antacid tablets.

Open food means food not enclosed and protected from contamination.

Part II General requirements

6 No food business permitted in insanitary premises.

7 All articles coming into contact with food shall be clean.

7 (1) Food containers must be cleanable and clean. They must be made of materials which prevent any injurious matter being absorbed.
The containers must prevent risk of contamination.

8 (a) Food business must not allow any food to be taken out to be prepared or packed in domestic premises.

Part III Protection of food from contamination

Food handlers must ensure that food is protected from risk of contamination:

- do not place food where it is likely to be contaminated.
- before offering any food for sale, separate it from any food unfit for human consumption.
- do not place any food in a yard, lower than 45cm (18in) from the ground, unless the food is enclosed and protected from contamination.
- cover or screen open food while it is exposed for sale and during delivery.
- do not have open animal food in a food room.

Personal cleanliness

Whilst handling food:

- keep clean any part of the body likely to be in contact with food.
- Keep clothing that is likely to be in contact with food clean.
- cover any cut or abrasion with a waterproof dressing.
- do not spit
- do not smoke or take snuff.

11 Personnel shall wear clean and washable overclothing while handling open food.

Wrapping of food

12 (a) No live animal or poultry must come into contact with food.
(b) Only film or paper designed for the purpose may be used to wrap food, no newspaper allowed except for wrapping:

- uncooked vegetables
- unskinned hare or rabbits
- unplucked game or poultry.

Reporting infection

13 Any food handler must tell his employer if he suffers from, or is a carrier of:

- typhoid
- paratyphoid
- *Salmonella*
- amoebic or bacillary dysentery
- Staphylococcal infection.

The employer must inform the Chief Environmental Health Officer of the infection.

Part IV Food premises

14 No direct air connection is allowed between a soil drainage system and the ventilation of a food room.

15 No cistern supplying water to a food room shall also supply a sanitary convenience other than through a flushing system which protects the water from contamination.

16 Sanitary conveniences used in food premises:
(a) must be kept clean and in good working order.
(b) be placed so that no offensive odours can pass into food rooms.
(2) The sanitary rooms must be well lit, ventilated and kept clean.
(3) A room containing a sanitary convenience must not be used as a food room.
(4) Open food must not be handled in any room directly connected with a sanitary room.
(5) A 'Now wash your hands' notice must be fixed in a prominent place near every sanitary convenience used by food handlers.

17 Water supplied to a food room must be clean and wholesome.

18 (1) There must be sufficient conveniently placed wash hand basins for the use of food handlers.
(2) The basins must have hot and cold water or water at a controlled temperature if open food is to be handled.
(3) Each basin must have soap or detergent, nail brush and clean towels or drying facilities.
(4) Basins must be kept clean and in working order.
(5) Hand basins must only be used for personal washing and for no other purpose.

First aid

19 First aid material must be readily accessible to all food handlers and include a supply of bandages and dressings (Including waterproof dressings).

Accommodation for outdoor clothing

20 Accommodation must be provided for food handlers' outdoor clothing. If in a food room, lockers or cupboards must be provided.

Washing food and equipment

21 (1) Sufficient sinks must be supplied for washing food and equipment.

(2) Sinks must be supplied with hot and cold water unless used only for:

(a) washing fish, fruit or vegetables
(b) or for ice-cream formers or servers or drinking vessels, when cold water may be used with a suitable bacteriocidal agent.

(3) Sinks must be kept clean and in working order.

22 All food rooms must be adequately lit.

23 All food rooms must be well ventilated.

24 No food room or adjacent room may be used for sleeping in.

25 The materials and structure of a food room must be cleanable and kept clean and in good repair so as to prevent as far as is possible, the entry of birds, rodents and insects.

27 (1) Certain types of foods, intended for human consumption must be kept at specified temperatures.
These are :

(a) soft cheese (whether cut or whole) which have been ripened by microbial action.
(b) cooked products (whether ready for consumption or intended to be reheated or cooked before consumption) containing:
  (i) meat
  (ii) fish
  (iii) eggs
  (iv) substitutes for meat, fish or eggs.
  (v) cheese
  (vi) cereals
  (vii) pulses
  (viii) vegetables (whether the food also contains other raw or partially cooked ingredients).
(c) smoked or cured fish
(d) cut or sliced smoked or cured meat
(e) desserts containing milk or milk substitutes having a pH value of 4.5 or more
(f) prepared vegetables salads, including those containing fruit
(g) cooked pies and pastries containing meat, fish or any substitutes for meat and fish or vegetables
(h) cooked sausage rolls except those to be consumed on the day of preparation or the following day
(i) uncooked or partly cooked pastry containing meat, fish or substitutes for meat and fish
(j) sandwiches and filled rolls containing:
  (i) meat
  (ii) fish

(iii) eggs
(iv) substitutes for meat, fish or eggs,
(v) soft cheeses
(vi) vegetables
(k) cream cakes

27 (2) This regulation does not apply to any of the following foods:
(a) Bread, biscuits, cakes or pastry – goods made safe by the high temperature of baking.
(b) ice-cream – covered by the Ice-cream (Heat Treatment) Regulations 1959.
(c) dehydrated (dried) foods treated to prevent the growth of pathogens
(d) uncooked bacon, uncooked ham, dry pasta, dry powders for making puddings or beverages
(f) chocolate or sugar confectionery
(g) milk is not combined with other ingredients

27 (3) This regulation does not apply to relevant foods in the following circumstances:
(a) if they are in a food room and are intended to be sold within 2 hours of the conclusion of preparation and are kept at 63°C or above.
(b) if they are in a food room and are intended to be sold within 4 hours of the conclusion of preparation and are kept at a temperature below 63°C.
(c) if the food is on display in catering premises for not more than 4 hours for either:
(i) indicating to purchasers the type of food on sale or
(ii) for service to customers on the premises.

27 (5) The high risk foods listed in **27**(1) if not among the exceptions listed above must be either:

- *hot* – kept at or above 63°C or
- *cold* – cooled as soon as possible to 8°C.

The lower specified temperature applies to *all relevant foods until 1st April 1993*.

On 1st April 1993 the foods listed below in list (9) will have to be kept at 5°C while all other relevant foods may still be kept at 8°C.

The foods which will have to be kept at 5°C on and after 1st April are:

(9)(a) cheeses which have been cut from the whole cheese
(b) foods listed in 27.1(b) which have been prepared for consumption and do not require heating or cooking, for example cold meats
(c) smoked or cured fish
(d) smoked or cured meat, sliced after smoking or curing

(e) sandwiches and filled rolls containing foods mentioned in this list unless intended to be sold within 24 hours of their preparation.

(7) the specified temperatures, (63°C and 5°C/8°C) may be exceeded by 2°C for not more than 2 hours under the following circumstances:
(a) during preparation of foods
(b) while defrosting foods
(c) during temporary breakdown of equipment
(d) during movement of food from one part of the premises to another

Regulation 27 may seem complicated but the aim is simple – to prevent the multiplication of spoilage and pathogenic organisms in vulnerable foods by directing them to be held, hot above 63°C or cold at 8°C/5°C.

The foods in paragraph (9) of the regulation which have to be kept at 5°C from April 1st 1990 are those most likely to contain organisms which grow readily at cool temperatures – between 5°C and 8°C.

Health and Safety at Work Act 1979 and food production premises

This Act requires *all employers* to provide a safe and healthy working environment for their employees. This includes the safety of all machinery used in the preparation and production of food and the provision of materials necessary for effective cleaning operations.

All employees are required to carry out their duties in a safe and responsible manner to ensure that they do not hinder or wilfully prevent the employer from providing a safe and healthy workplace.

A summary

Employers' duties

The employer must, so far as is practicable, ensure that:

1 machinery, equipment and other plant provided is safe and without risk to health and is maintained in a safe working condition.

2 working systems are safe and without risk to health by:

- correct layout of workplace
- systematic organisation of work
- laying down safe work schedules
- specifying precautions required before carrying out certain tasks

For example, machinery must be correctly assembled; correct operating procedures must be known; safety notices must be displayed.

3 the health and safety employees are not at risk when handling dangerous materials or when these materials are in transport or in storage. Such materials include, for example, cleaning materials, cleaning chemicals and gas bottles.

4 all employees are provided with the information, training and supervision necessary to ensure their health and safety at work. The information must include:

- hazards of the workplace and how to avoid them e.g. slippery stairs, floors
- spillage of liquids or grease on the floor
- notices for machinery and operating instructions
- regulations and codes of practice.

5 training includes safety instructions and emergency procedures, for example:

- regular servicing of machinery and power sockets
- fire drill
- first aid procedures
- retraining, when the introduction of new machinery requires staff to be instructed in new operation and safety procedures.

6 managers and supervisors are made fully aware of their responsibilities:

(a) to ensure that places under their control are kept in a safe condition:

- the floor is in a good state of repair and clean
- entrance and exits are in good repair
- entrances and exits are well lit
- stairs and passages are well lit

(b) to ensure that the working environment is safe and clean and satisfactory as regards:

- heating, lighting and ventilation
- washing facilities
- changing areas and clothes lockers.

Under the Act employers must prepare a written policy statement of the safety policy and all staff must be given a copy or be made aware of it, by the policy statement being displayed in strategic positions e.g. in kitchens. The statement must list persons responsible for specific areas e.g. chef de cuisine.

Employees' duties

Employees must take reasonable responsibility for their own health and safety. They also have a duty of care for their colleagues who may be affected by what they do or neglect to do. Employees must:

- avoid silly or careless behaviour
- avoid putting others at risk by their actions, e.g. wipe up spillage on floors *immediately*

REPORT FORM – ACCIDENTS TO EMPLOYEES

This form must be completed in the case of all accidents to employees

DATE: 15 October 19–

INJURED PERSON

Full Name (block capitals) JOHN WILLIAM SMITH Age 19

Home Address 7 Cherry Tree Rise, KENDAL

Occupation COMMIS CHEF

Married? Single? ✓ (Please tick as appropriate)
Male? ✓ Female?

Date started employment 21 June 19–

PARTICULARS OF ACCIDENT

Date 15 Oct Time 11.15 Place KITCHEN

How did the accident occur? (Please give details of any equipment involved and continue overleaf with diagram if helpful). The casualty was preparing parsnips with a knife when the knife slipped

Injuries sustained Deep cut to the thumb of the left hand

Witnesses (if any) (1) Mary Brown Occupation Stores Assistant

(2) Occupation

Give details of any safety clothing worn The casualty was correctly dressed.

Was the employee authorised to be in the area of the accident? Yes ✓ No N/A

What action has been taken to prevent reocurrence of the accident? Guidance on the correct method to prepare vegetables

Was the necessary training given Yes ✓ No N/A

Has the employee ceased work? Yes No ✓ if so on what date?

Has the employee returned to work? Yes ✓ No if so (a) on what date?

(b) at normal duties Yes ✓ No (c) at normal pay? Yes No

SIGNED J. Bloggs Signature (First Aider)

J BLOGGS Name (block capitals)

SIGNED W. A Milne Signature (Manager or Supervisor)

W. A. MILNE Name (block capitals)

This form must be submitted as soon as possible following the accident. Please note all accidents should be recorded in the accident book.

00-619-4

Fig 1.7 Accident report form

- keep all gangways and floors free of obstructions
- see that machinery is reassembled and tested before leaving it
- not run in areas where machinery is kept or equipment is in use
- carry knives correctly
- ensure that colleagues are aware of temporary hazards
- report all defects or substandard conditions *immediately* to the person responsible, e.g. broken power sockets, faulty appliances, poor lighting, ineffective machine guards, etc.

First aid There must be at least one first aid box in a kitchen and the box must be fully maintained with the correct dressings and plasters as laid down by the Health and Safety at Work Act. The first aid box must be situated where it is easily accessible to all staff members. There must be a full time member of staff trained in first aid on each duty shift unless a medical supervisor is on site. The names of the first aiders should be displayed in a prominent position.

Any injury sustained by an employee which requires further hospital treatment must be recorded in an accident book according to the safety policy. This record will bring to light any defective equipment or bad practice which should be investigated.

First Aid kit requirements

- 3 packs of 14 sterile dressings (coloured)
- 8 sterile medium dressings (no8)
- 4 sterile large dressings (no9)
- 4 sterile extra large dressings (no3)
- 4 sterile large pads (no16)
- 4 sterile triangular bandages
- 2 bunches of 6 safety pins
- 100g pack of cotton wool
- 2 or 3 melolin dressings
- 1 micropore surgical tape
- 3 pairs disposable gloves
- 3 disposable bags
- 1 plastic apron
- 1 leaflet on advice for First Aid treatment

Exercise

1. Name the three Acts which affect the operation of all food production premises and food handlers.
2. State who is responsible for carrying out the requirements of these Acts.
3. State the personal responsibilities of the food handler under this legislation.
4. List the items which should be supplied in a kitchen first aid box.

2 Basic knife skills

All chefs have their own sets of knives which they look after. They will very rarely lend their knives to others because each individual uses them to cut at a different angle. The same knife can be sharp to one chef and seem blunt to another. Chefs' or cooks' knives are very expensive to replace so you must look after them and keep them in good condition.

Use and care of knives

Basic skills are acquired by careful attention to expert tuition but these skills need to be improved by constant practice. You need to be made aware of the dangers of incorrect procedures whilst you are developing your skills and be discouraged from taking short cuts which could lead to accidents.

The most widely used basic skill of the chef or cook is knife drill. Knives are the main tools used in food preparation and are available in many shapes and sizes. They are designed for specific purposes and should, therefore, be used *only* for those purposes.

The style of knife most widely used by the chef or cook is the French Cook's pattern, which is available in various sizes. The handles are shaped to fit comfortably into the hand and modern designs have plastic handles which have a slightly rough surface for maximum grip. These plastics handles are also more hygienic than the older wooden type. Knives are available with handles colour coded for use with particular foods, e.g. raw meat, fish and cooked meats.

To hold your knife correctly and safely, first make sure that your hands are free from grease or moisture so that you have maximum control. Pick up the knife so that the thumb of the hand holding the knife is along the flat of the handle, facing you, and the index finger rests on the top of the handle and passes over to the other side. The three remaining fingers are used to grip the handle underneath. Hold the handle firmly so it feels comfortable but do not grip it too tightly as this will cause tiredness or cramp in the hand or arm.

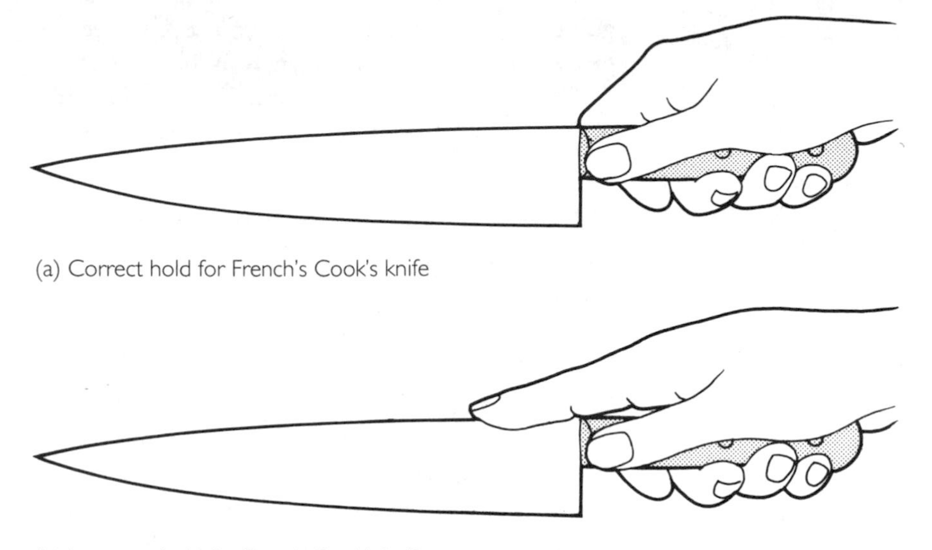

(a) Correct hold for French's Cook's knife

(b) Incorrect hold for French Cook's knife

Fig 2.1 Correct and incorrect ways of holding a French Cook's knife

The tip or point end of the knife is known as the toe and the wider, thicker end as the heel, useful terms to know when discussing cutting methods.

Your large chopping knives may be used for cutting, using two different basic movements.

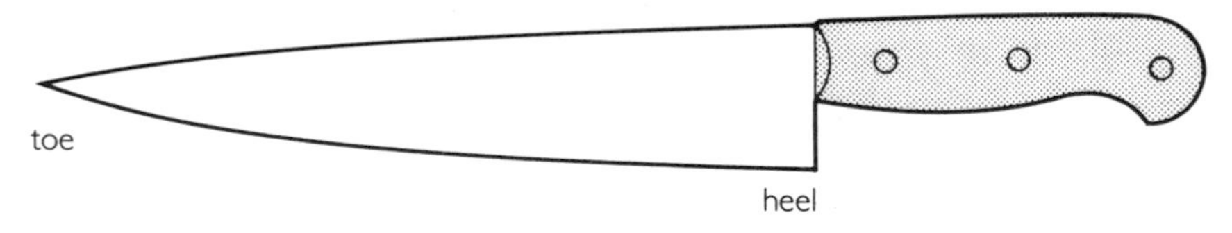

Fig 2.2 Chopping knife showing toe and heel

Method A

1 Place the toe of the knife on the cutting surface and holding the handle as described above, push forward and down at the same time.

2 Lift the handle and repeat the sawing or rocking action to cut the food. The toe remains in contact with the food.

3 The fingers of your other hand, hold the food with the finger tips turned slightly under to prevent them being cut. The knuckles serve as a guide to gauge the thickness of the cut (*see* Fig 2.4(e)). As the cutting process proceeds, your fingers walk backwards.

This method is used for shredding, slicing and cutting various food, and is a skill you can soon master with practice.

Task

Practise using your knife as described and using your finger as a guide.

Method B

With this method, you pull the blade of the knife down and towards the food and at the end of the pull, the knife should be in contact with the board. The finger of the other hand is used as already described.

This method is used for slicing cucumbers, tomatoes, carrots and onions. When dicing onions, both of these movements are used.

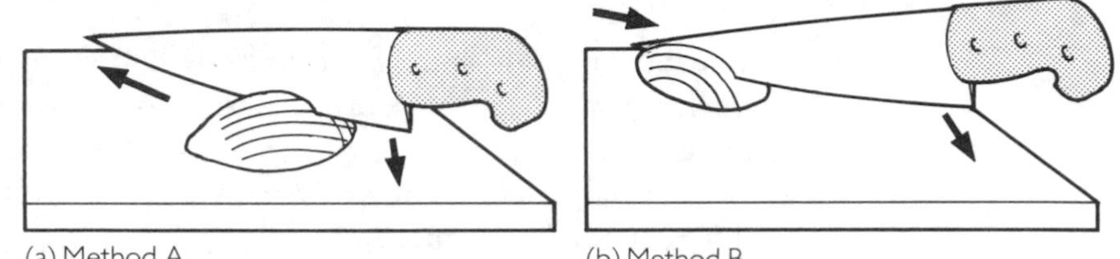

(a) Method A (b) Method B

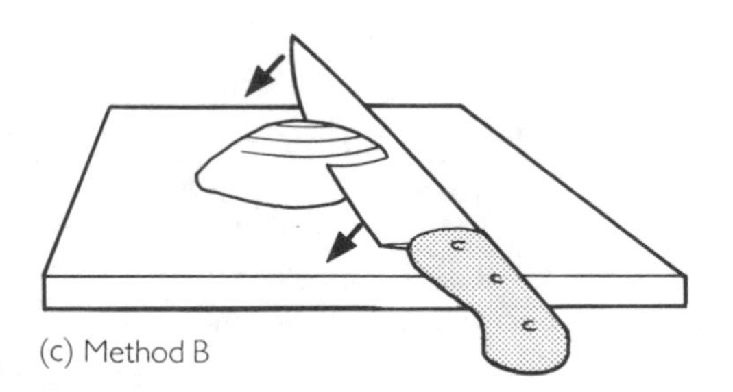

(c) Method B

Fig 2.3 Two methods of cutting an onion

1 Diagram (b) Method B Pulling knife down and towards the food, making vertical cuts.
2 Diagram (c) Method B Knife held in horizontal position pulling knife towards the food.
3 Diagram (a) Method A Knife with toe on the board.

Method B is used for cutting vertical and horizontal slices and method A for cutting the dice.

Most of your knives will be held in the way described above but with slight variation in grip.

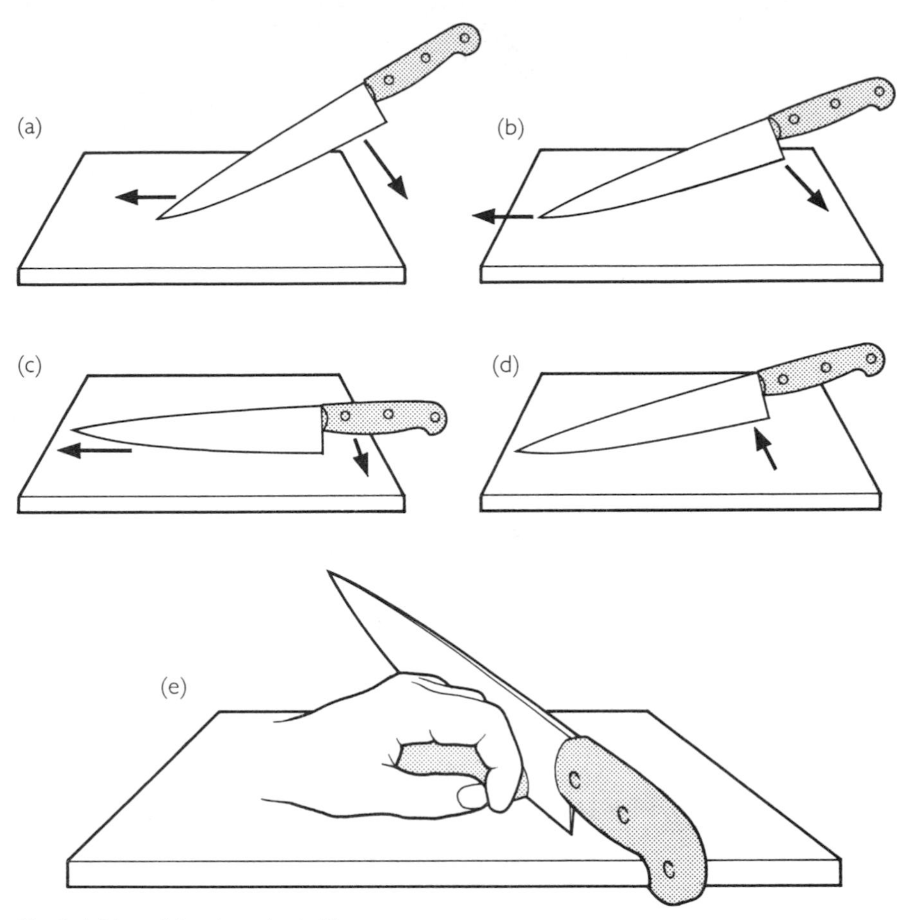

Fig 2.4 Use of the chopping knife

Method of cutting with knives

Having practised holding your knives and getting used to the feel of them, you are ready to start cutting food.

Food must always be prepared before cutting e.g. vegetables washed, cuts of meat skinned or defatted, fruits washed and fish washed or gutted. Your working surface must be clean and clear of unwanted equipment.

Work flow is important when cutting foods, whether in large or small quantities:

- The food *to be cut* should be placed to your *left.*
- Stand at right angles to the table, with your weight distributed evenly on both feet.
- The food, when *being cut* should be in *front* of you.
- *After cutting*, it should be placed to your *right* on the board or tray. It is important to **clear as you work.**

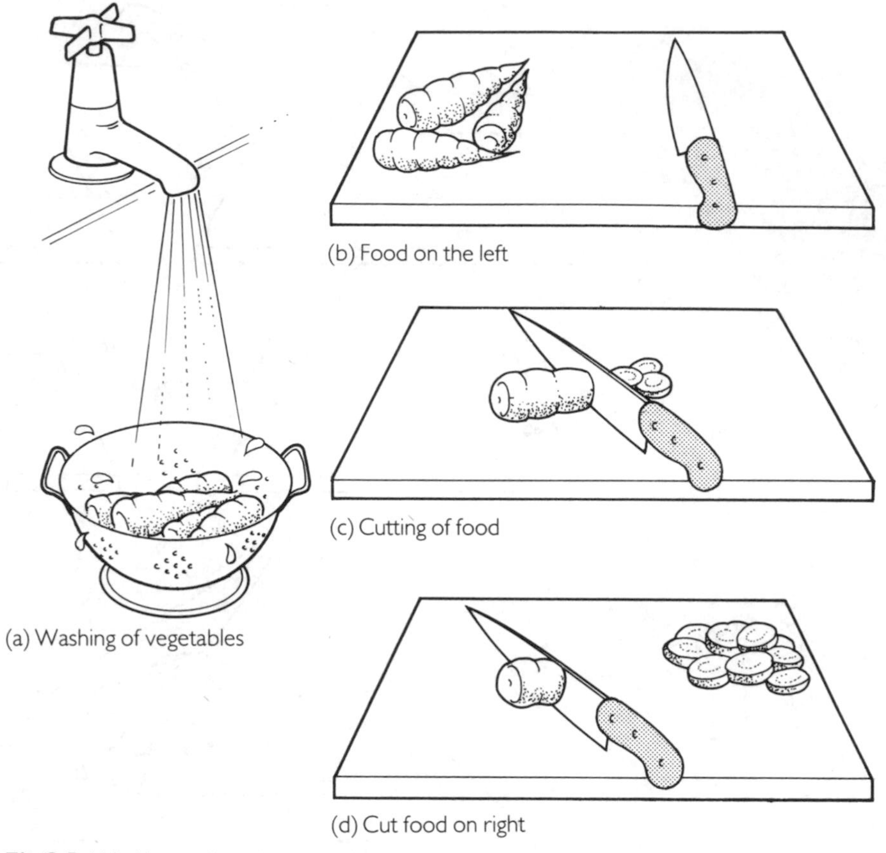

(a) Washing of vegetables

(b) Food on the left

(c) Cutting of food

(d) Cut food on right

Fig 2.5 Washing and cutting vegetables

Characteristics of various knives

Office knife

Short, stout blade 10.5 cm (4in) in length with a 10.5 cm (4in) handle. It is used for preparing vegetables e.g. peeling onions, preparing Brussels sprouts, eyeing tomatoes and turning vegetables.

The knife is held comfortably in the hand and the short blade and good sized handle ensure comfort and full control in use.

Paring and turning knives

Paring knives have rounded tips to the blades and are of similar length to office knives.

Turning knives have curved tips and are of a similar length. These two knives are used for turning vegetables into barrel shapes e.g. potatoes, carrots. They are held in a similar way to the chopping knife but the thumb supports the food being turned, giving greater control of the knife.

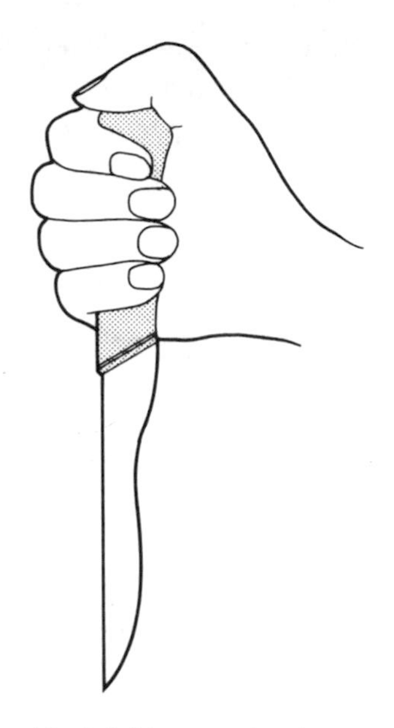

Fig 2.6 How to grip a boning knife

Boning knife

Strong, stout blade which varies in length, a 15 cm (6in) blade being the most widely used. The blade curves to a point onto the straight back. The handle is shaped so that it fits comfortably into the hand when held like a dagger. Held this way, the point is in contact with the bone. The blade remains rigid when pressure is applied. This knife is used for meat and is *unsuitable* for any other purpose. Attempts to use it for other jobs are *dangerous*.

Filleting knife

A narrow 18 cm (7in) flexible blade with 10.5 cm (4in) handle. The flexible blade allows the fillets of fish to be removed from the bone, cleanly and without cutting the flesh. If this knife is used for cutting food, the blade will bend when pressure is applied and perhaps cause an accident.

Palette knife

A very flexible blade with a rounded tip. The blade is available in various lengths and widths. It is used for removing baked foods from trays, spreading various mixtures and coating cakes and sponges with creams and icings. **It must never be used for cutting.**

It is held as previously described but the wrist moves through 180°C according to use.

There are many other knives designed for specific purposes which you may use in your career, but always remember to **use the correct knife for the job!** Treat knives with the respect that they deserve and they will be good friends, abuse them and they will become enemies.

Sharpening your knives

During your working day, your knives will lose their cutting edges and this can be remedied by using a butcher's steel.

The butcher's steel has many grooves along its length which provide an abrasive surface. They can be obtained in various lengths but always make sure you have a steel long enough to sharpen your longest knife. The grooves become clogged with grease after a period of use so the steel must be washed thoroughly to remove it and restore the abrasive surface.

There are two methods of sharpening knives. Try them both and select the one you prefer and find safer to use.

Method A

Hold your steel firmly in your left hand and the knife to be sharpened in the right hand.

1 Place the heel of the knife to the tip of the steel on the side facing you, with the knife at an angle of approximately 30°.

2 With a wrist movement, bring the blade down the length of the steel, keeping the toe of the knife to the tip of the steel.

3 Repeat on the other side of the steel to give five strikes on each side of the knife. This should be sufficient to resharpen, if it is done regularly during your work.

Method B

1 Place the point of your steel on a folded cloth, to prevent it slipping on the table.

2 Place the *heel* of your knife to the *guard* end of the steel and draw the length of the blade across the steel so that the whole blade is sharpened.

3 Repeat as before.

After sharpening, **always wipe the blade** to remove grease and fine metal turnings which have adhered to your knives.

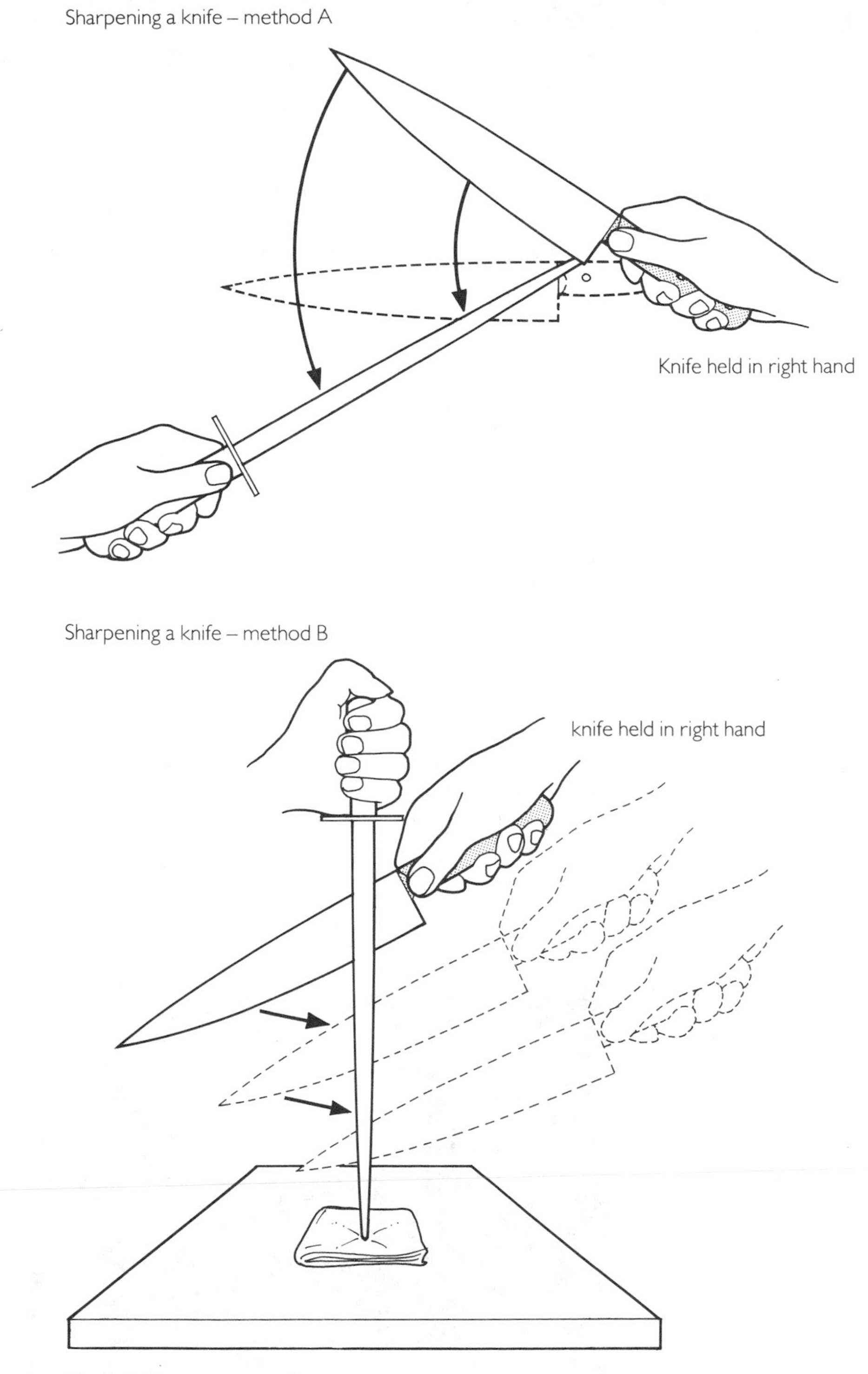

Fig 2.7 Sharpening a knife

Task

When you sharpen your knife, wipe the blade on a clean cloth. You will be surprised at the dirt left on the cloth.

Sharpening the whole length of the blades as described above will ensure that they keep an even cutting edge. If you have only sharpened parts of the blade, wear will result in the cutting edge being uneven and the food will not be cut properly.

Exercise

1. What are the important points to observe before you start to use your knives?
2. Describe the characteristics of the following knives and state their specific uses: **(a)** filetting, **(b)** boning, **(c)** turning, **(d)** chopping, **(e)** palette.
3. Describe a butcher's steel and state its use.
4. Why is it necessary to wash or wipe your knives after sharpening them on a steel?
5. What would happen to your knife blade if you sharpened it incorrectly?

Care of knives

Your knives are used extensively for cutting numerous foods, both raw and cooked. Therefore you must take special care to wash them in hot water and detergent, rinse and dry them after each time they are used for cutting different foods to avoid cross contamination.

Fig 2.8(a) Correct storage of knives

Fig 2.8(b) Incorrect storage of knives

When washing knives, pay particular attention to the area where the blades and the handles meet. Food becomes trapped in the crevices, particularly fatty and greasy foods such as hot roast meats, cold meats, raw meat and fish. Knives must always be washed and dried thoroughly before they are put away. Many of your knives may not be stainless steel and so will rust if put away slightly damp or with food particles on them. Rusty or dirty knives are dangerous because they contaminate foods.

Always store your knives in a safe manner i.e. in a knife case, knife box or drawer kept specifically for this purpose. Avoid just throwing knives into a drawer, as this is dangerous and will spoil their cutting blades.

Points to observe when using knives

Do's

- Make sure knives are laid with the cutting edge flat on the table and not facing upwards.
- Wash and dry knives with the *back* of the blade towards the hand **never** the *front* of the blade.
- Keep knives sharp. Remember **blunt knives cause accidents**.
- Ensure that knives are clean and dry before putting them away.
- Use the correct knife for the purpose.
- Carry knives with the points towards the floor. It is safe and cannot cause injury to anybody.
- Place knives where they can be seen, when they need washing. Better still – always wash your own knives.
- Keep all knives correctly covered when not in use, either in a case, box or tidily in a drawer.

Don'ts

- Don't cover your working surface with knives which you are not going to use for your particular task.
- Never allow your knives to become covered with peelings, prepared or unprepared food, as you could cut yourself when clearing away the peelings or placing food on trays.
- Never attempt to catch a falling knife.
- Never place your knives in a sink when they need washing because the washers-up could cut their hands badly.
- Don't borrow or lend knives. You get used to your own tools. Borrowed tools may be difficult or uncomfortable to use and could cause accidents.
- Never stretch across or in front of anyone using a knife.

Exercise

1 How do you ensure that your knives are kept in working order?

2 What precautions must you take when washing and drying your knives?

3 List the points concerned with the safe use of knives.

3 Micro-organisms, control of spoilage and food poisoning

Good hygiene is important in food premises for three main reasons:

- to prevent waste of food materials
- to prevent food poisoning
- to fulfil legal obligations

Preventing waste

Vegetables will have a good shelf life if they are:

- selected carefully at the time of purchase
- stored correctly, in clean, airy racks
- inspected daily and any rotting or mouldy items, thrown away

However, vegetables will prove a bad buy if they are:

- bought carelessly, without inspection
- thrown in a box and bruised, damaging the protective outer skin
- not inspected daily, so infection passed from one vegetable to another

Note: Poor purchasing and improper storage cause high wastage and low profits.

In the preparation of fresh foods, there will always be some *unavoidable* waste for which the caterer has to budget – potatoes, for example, are peeled and the peelings thrown into the swill bin.

Exercise

List the type of waste in the preparation of – fresh peas, cauliflowers, meat, fish.

Good hygienic practices make sure that fresh foods have their correct shelf life and do not have to be thrown away because they have become rancid, rotten or mouldy. This prevents *avoidable* waste and protects profits.

Preventing food poisoning

Hygiene is important in all stages of food handling because foods (and drinks) can harbour dangerous micro-organisms which can cause illness in food handlers and their clients. You will want to known how to protect your own health and how to prevent your customers from suffering the unpleasant consequences of acquiring food poisoning after a meal at your establishment. If they do suffer in this way and can trace it back to your premises, they are not likely to favour you with their custom again and will no doubt tell their friends about their misfortune and your business will suffer.

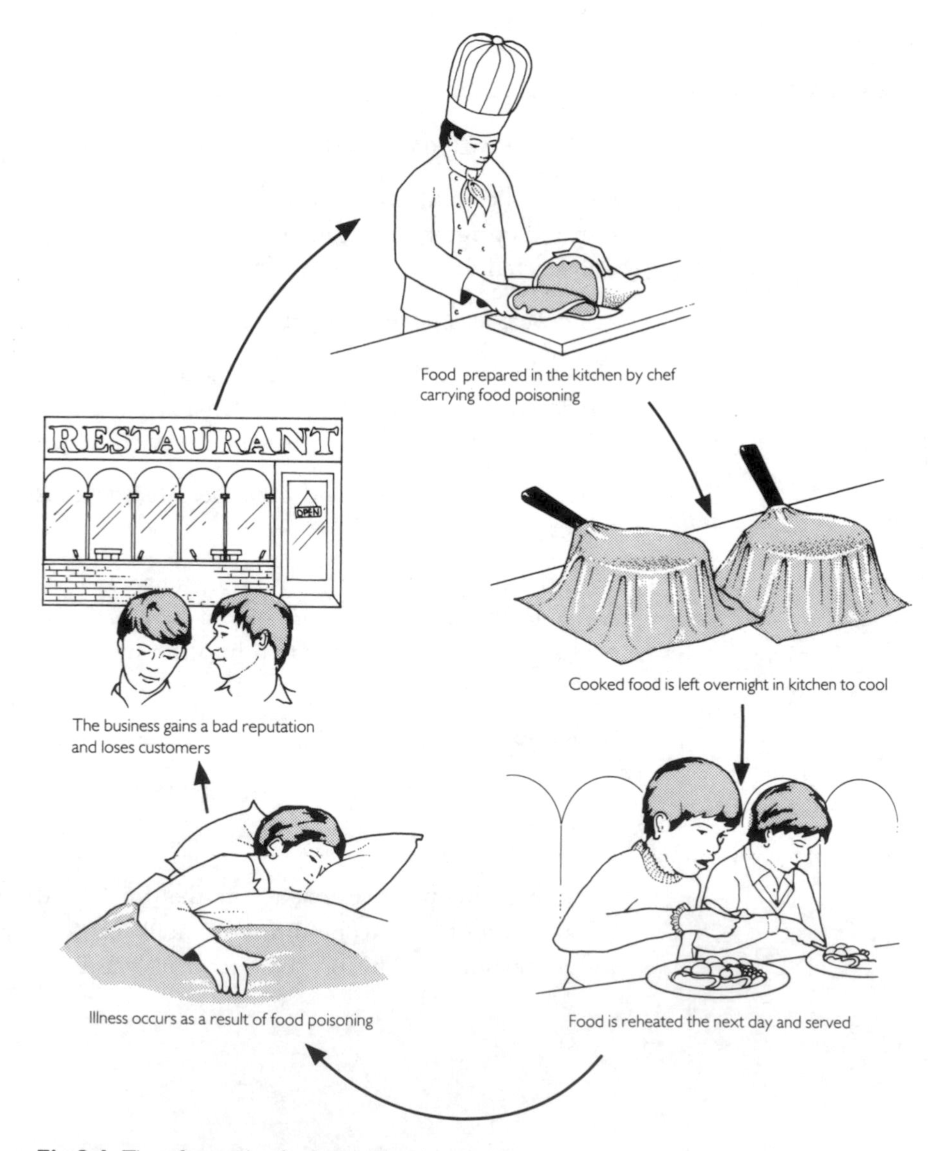

Fig 3.1 The aftermath of a food poisoning outbreak

Legal obligations

Lastly, as you saw in the previous chapter, good hygiene is a legal requirement in all food outlets. A food poisoning outbreak at your establishment, will at the least cause adverse publicity and may, if serious lapses of hygiene have occurred, lead to heavy fines and the possibility of closure of your business.

Micro-organisms

Micro-organisms are everywhere, in the air, on the surface of foods, on our clothes, on the surface of the skin and inside our bodies. The vast majority of these small organisms are *harmless*. They are part of the natural cycle of growth and decay. Without them, the world would be littered with bodies of dead animals and plants and the materials and energy in them would be locked away, unable to sustain further generations of living organisms.

Types of micro-organisms associated with food

There are four groups associated with foods:

1 Moulds
2 Yeasts
3 Bacteria
4 Viruses

These groups are concerned with food in a number of different ways:

- as **spoilage** organisms – causing deterioration in foods; rotting, mouldiness, putrefaction and rancidity.
- as **pathogens** – causing disease in man and animals
- as **useful** organisms – employed to produce or flavour foods

Moulds

Unlike other micro-organisms, the colonies of moulds are large enough to be obvious to the naked eye. When mould colonies are examined under a microscope, however, they are found to consist of long, thin filaments known as *hyphae*. The hyphae are of two kinds; *nutritive* hyphae which grow within the food and *reproductive* hyphae which grow upwards into the air.

The nutritive hyphae produce enzymes which make the nutrients in the food soluble, so the mould is able to absorb and use them. The changes brought about by mould growth are quite obvious; the firm structure of the food becomes soft and watery and the waste products of the mould give the food an unpleasant taste and smell.

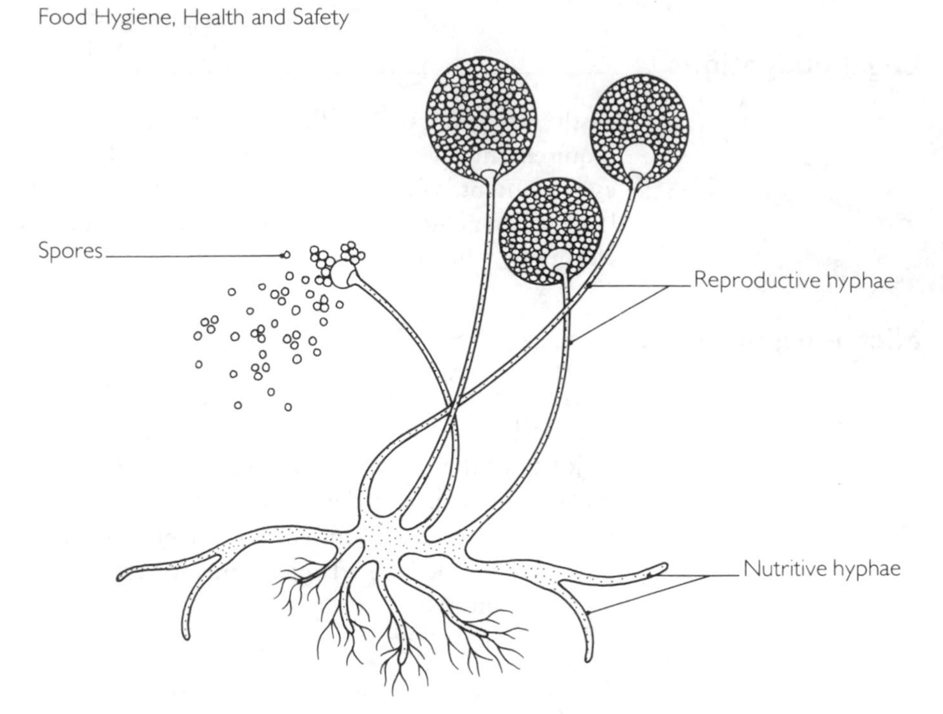

Fig 3.2 Nutrition and reproductive hyphae of *Rhizopus spp* and *Aspergillus*

The reproductive hyphae produce large quantities of *spores* which escape into the air. Mould spores are very light, so they can stay suspended in the air for long periods. If they reach a suitable moist food, the spores germinate and new hyphae emerge to start the cycle of spoilage once more.

Task

Set up a simple experiment using three petri dishes – one containing dry bread and covered, the second containing moist bread, covered, and the third, moist bread, uncovered.

Exercise

1 Which piece of bread grew an abundant crop of mould within a few days?
2 List the factors necessary for mould growth to occur.
3 Suggest two ways you can prevent mould spores from passing from one food to another.

Moulds are used in the *production* of some foods. Moulds of the *Penicillium* group are used to develop the characteristic flavours of a number of cheeses. Roquefort cheese, for example, is the result of adding *Penicillium roqueforti* to ewes milk cheese. To quicken the growth of mould, the cheese is pierced with long needles. In producing Camembert cheese, *Penicillium camemberti* is added – but applied to the surface of the cheese.

Task

Scrape a little of the white furry growth from the surface of a Camembert or Brie cheese. Mount in a drop of water and examine under a microscope. Can you find some of the hyphae and perhaps some spores of the mould?

Mouldy food

Mould attack is responsible for considerable wastage of food at all stages of production, from the farm to the kitchen, but it is rarely responsible for causing illness in man. However, some moulds are known to produce poisons known as *mycotoxins*. Some of these substances have caused serious diseases in animals and are suspected of occasionally affecting human beings. Some mycotoxins, known as *aflatoxins*, have been shown to be capable of causing cancers.

Note: In view of these findings, it is better to err on the side of caution. ***Throw away mouldy foods*** *rather than scooping off mould from the top of jam or cutting the mouldy crust from cake and using the* ***apparently*** *unaffected part.*

Yeasts

Yeasts belong to the same group of plants, the fungi, as the moulds but they consist of small, single cells. They are capable of fermenting sugars to produce carbon dioxide and alcohol. Carbon dioxide is used to raise bread and alcohol is the basis of the brewing industries. A large industry exists to produce pure cultures of yeasts for baking and brewing.

Wild yeasts can cause spoilage, particularly in sugary foods such as syrups and honey. Some yeasts form skins and 'off' flavours in beers, and other types spoil fatty foods such as butter and margarine.

Bacteria

Bacteria, like the moulds and yeasts mentioned before, are a major source of spoilage in foods. As they are widespread in soil, dust and water, they are likely to cause some form of spoilage – putrefaction in meats and fish, soft rots in fruit and vegetables, and rancidity in fats whenever these foods are left in the warm, moist conditions which favour their growth.

Food poisoning

Some groups of bacteria are *pathogenic* to man and some of these can *multiply* in food and drink, causing food poisoning symptoms if the food is eaten. Others do not multiply but can *survive* in food and drink – so pass food-borne diseases to the consumer.

Like yeasts, bacteria are single cell organisms, but much smaller in size. Even with 1000-1200X, the highest magnification obtainable under a light microscope, bacteria appear as small spheres, rods or curved shapes as shown in Figs 3.3 a, b and c.

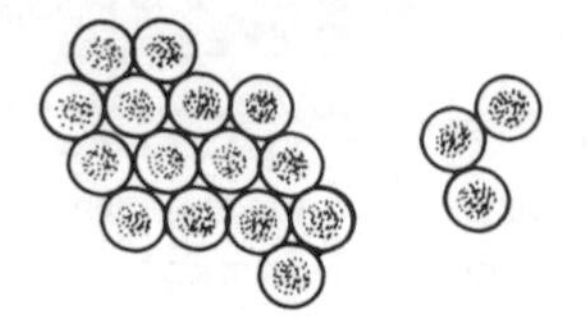

(a) Round cells – cocci Staphylococcus aureus

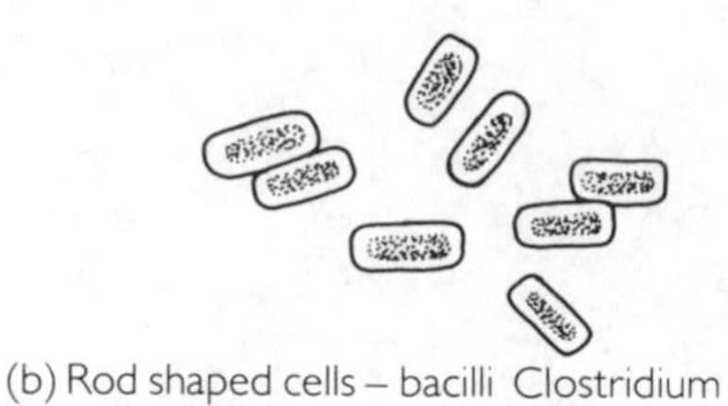

(b) Rod shaped cells – bacilli Clostridium perfringens

(c) Curved and comma shaped cells – Vibrio cholerae

(d) Vibros and campylobacters – Campylobacter jejuni

Fig 3.3 Shapes of bacterial cells
(a) Spherical cells *(Staphylococcus aureus)*
(b) Rod-shaped bacilli *(Clostridium perfringens)*
(c) Comma shaped cells *(Vibrio cholera)*
(d) Curved cells *(Campylobacter jejuni)*

Reproduction

Under favourable conditions, bacteria multiply extremely rapidly. They do so very simply, by dividing into two, a process known as *binary fission*.

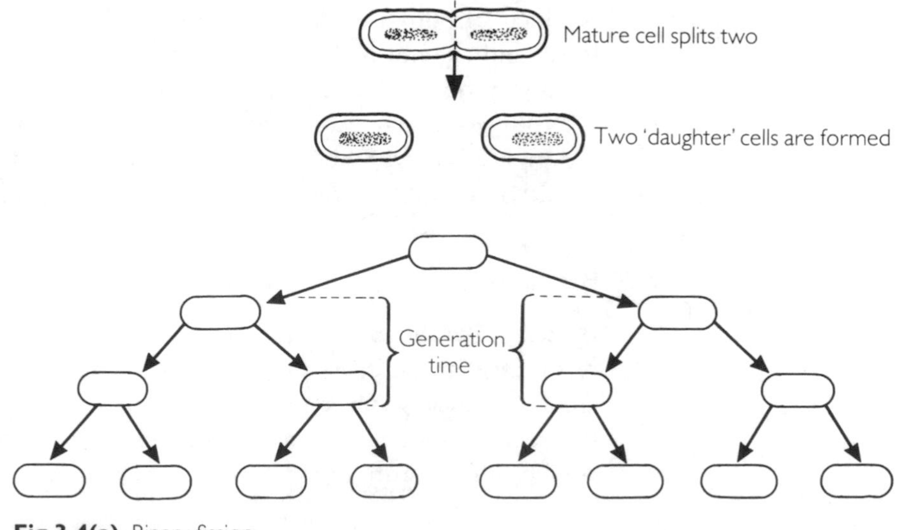

Fig 3.4(a) Binary fission
(b) Binary fission and generation time for bacteria

The time needed to double the number of bacteria, known as the *generation time*, varies with the amount of food available, the temperature, and the type of bacteria concerned. In the most favourable conditions, the majority of bacteria will divide into two within twenty minutes. At this rate of doubling, a single cell could produce more than 8000 descendents in four hours!

Task

If conditions were made less favourable for the bacteria mentioned above, so that the generation time was 40 minutes instead of 20, how many bacteria could be produced from a single one in 4 hours?

Vegetative bacteria

When bacteria are growing actively as in Fig. 3.5, they are said to be in the *vegetative* state. Vegetative cells are easily killed by heat – especially *moist* heat. Two minutes at boiling point is enough to destroy vegetative cells of bacteria. They can also be killed at temperatures between 65°C and 100°C if more time is allowed. For example, if you cook a joint of meat long enough for the centre of the joint to reach 75°C, the vegetative bacteria will be killed.

Sporing bacteria

Some groups of bacteria are capable of producing *spores* when conditions are unsuitable for active growth. Bacterial spores are small, round bodies with thick walls which allow the organism to stay dormant, but alive, over long periods of time. When conditions improve, the spores germinate and the bacteria return to the vegetative state and multiply rapidly once more.

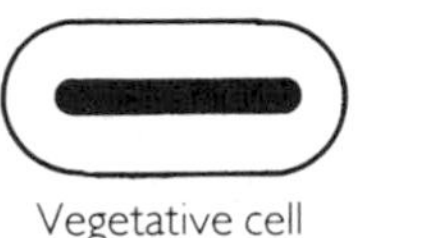

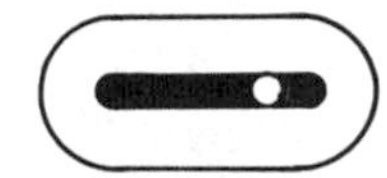

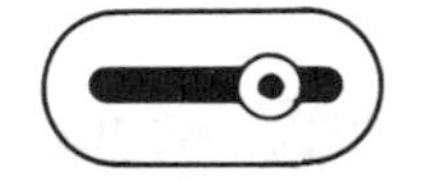

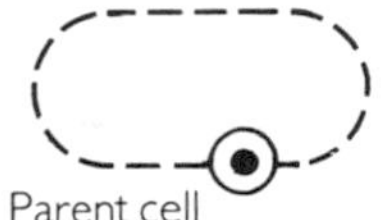

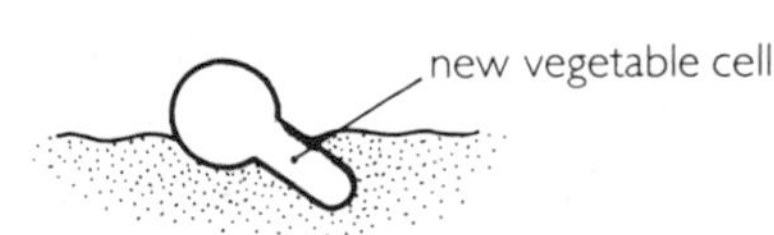

Fig 3.5(a) Formation of bacterial spore
(b) Germination

Spores are difficult to kill. Some can survive boiling for up to four hours, so low temperature methods of cooking such as poaching, braising and stewing do **not** kill spores. Spores will also survive treatments with chemical disinfectants which will kill vegetative bacteria.

Positive uses of bacteria in food production

A number of bacterial species are employed in the production of food and drink. Lactic bacteria are used to ripen some butters and in the production of cheese, yoghurt and fermented milk drinks. They also produce the preservative acids in cucumber and cabbage pickles and are involved in the processing of green olives. Acetic acid bacteria are used to sour fruit juices, cereals or alcoholic drinks to make the vinegar used in other pickles.

Viruses

Viruses are different from all other living organisms in that they do not consist of cells. They lack, therefore, the complex systems which allow other organisms to eat, respire, excrete, grow and react to the environment around them. In essence, they consist of a simple reproductive apparatus wrapped up in a protective envelope. Viruses can only multiply *within living cells*. They take over the systems of the cells they invade and instruct them to make more virus particles. They are *very small* – so small that they cannot be seen with even the highest power of the light microscope. It is only since the advent of the electron microscope, giving a magnification of 15 000X, that it has been possible to photograph individual virus particles.

A number of viruses, pathogenic to man, are known to persist in food and drink. Amongst these are the enteroviruses which cause symptoms of diarrhoea and vomiting similar to bacterial food poisoning and the viruses which cause polio (*poliomyelitis*) and liver infections (*hepatitis*). Little is known about how viruses are transferred by food and drink but it seems that methods of cooking which kill bacteria also kill viruses. The majority of outbreaks of viral diseases traced to food or drink have occurred:

- in uncooked foods e.g. raw milk
- in shellfish raised in polluted waters
- through contamination by a human carrier of the disease.

Summary

- Good hygiene is a legal requirement in food premises.
- Hygienic conditions control waste of food materials.
- Hygienic conditions control food poisoning and food-borne diseases.
- Moulds, yeasts, bacteria and viruses are the microbes associated with foods.
- Spoilage organisms cause deterioration of food.
- Pathogens cause disease in man and animals.
- Some types of micro-organisms are used in the production of foods.
- The spores of moulds spread infection from one food to another.
- Some species of moulds produce mycotoxins which cause illness in animals and possibly in man.
- Food poisoning organisms *multiply* in foods.
- Bacteria causing food-borne diseases *survive* in food and drink.
- Bacteria multiply by binary fission, that is by dividing into two.
- Bacteria which are actively dividing are in the vegetative state.
- Vegetative cells are easily killed by heat and chemical disinfectants.
- Spores are dormant bodies and are resistant to heat and chemical disinfectants.
- Viruses are non-cellular.
- Viruses can only multiply within living cells.

Self assessment exercise

1. Explain what is meant by the terms (a) pathogen and (b) spoilage organism.
2. Name two cheeses whose characteristic flavours are due to mould growth.
3. Which industries are based on the ability of yeasts to ferment sugars?
4. What action should you take if you found a food was mouldy? Give two reasons why you must take the action that you have described.
5. Name two food products made as a result of bacterial action.
6. Explain the difference between food poisoning and food-borne disease.
7. Describe the way in which bacteria multiply. What name is given to this process?
8. What is meant by the generation time?
9. If the generation time is shortened, what effect does this have on the number of bacteria produced per hour?
10. Which of the following descriptions fits a bacterial *spore?*
    (a) A light, reproductive body, capable of travelling in air currents.
    (b) A dormant body, capable of resisting adverse conditions.
11. What is meant by bacteria being in the vegetative state?
12. Name one method of cooking which will kill vegetative pathogens but not their spores.
13. State two ways viruses differ from bacteria.
14. Describe one way a food might become contaminated by a pathogenic virus.

Control of growth of micro-organisms

Six main factors control the growth of micro-organisms:

- Food
- Water
- Temperature
- Oxygen
- Degree of acidity/alkalinity
- Time

If we understand these factors, we can use them to *prevent* the growth of food poisoning or spoilage organisms. Alternatively, if we wish to use micro-organisms to make foods, we can use these factors to *create* optimum conditions for their growth.

Food

Like all living things, microbes need food for growth and energy production. Food is the most difficult factor to control since, if foods are nutritious for human beings, they will also sustain micro-organisms.

However, there are some measures that can be taken to protect foods from microbial growth. Manufacturers can add preservatives – chemical substances that slow the growth of microbes but are safe for people to consume when used at low concentrations. As regards fresh produce, all caterers can do is to buy clean, good quality food and protect it from contamination during storage, preparation and cooking.

Water

Water is vital to all life – no micro-organism can grow or multiply without a supply of water in the *liquid* form. The simplest way of denying microbes the water essential for growth is to *dry* the food. Foods such as pasta, dried pulses, milk and egg powders will not support the growth of spoilage or food poisoning organisms providing they are *kept* dry. If left exposed to the air, moisture will be absorbed and the micro-organisms will be able to grow.

Water can be *present* in a food but *unavailable* to micro-organisms. The high content of sugar in preserves (jams, syrups, etc.) or of salt in such foods as bacon and kippers, makes the water in these foods unavailable to micro-organisms. The sugar or salt withdraws water from the microbes by a process known as *osmosis*, effectively dehydrating them and preventing them from growing.

Task

Cut some thin slices of cucumber and spread them out on a plate. Sprinkle the slices with salt. Examine them one hour later.

Exercise

1 What change did you notice in the appearance of the cucumber slices?

2 Name the process responsible for the change.

3 If there had been organisms in the cucumber, what effect would the salt have had on them?

Freezing When a food is frozen, the water becomes 'locked up' in the form of ice and so is not available for micro-organisms. At the same time, the low temperature needed to keep the food frozen is an extra protection against spoilage or growth of food poisoning bacteria.

Temperature

Bacteria grow best in *warm* conditions. The danger zone for rapid multiplication of bacteria is between 10°C(50F) and 63°C(145F). Above 63°C they cease to multiply. It is for this reason that the Food Hygiene

Regulations 1970 instructed caterers to hold vulnerable foods cold (below 10°C) or hot (above 63°C). However current practice is to hold foods below 4°C or above 70°C, for extra safety. Pathogenic organisms grow most rapidly at about *blood heat* – 35°C to 37°C. However, the optimum growth temperature for most spoilage organisms is between 25°C to 30°C – temperatures which are easily reached in storerooms on hot summer days.

Heating foods

Increasing the temperature of a food above 65°C quickly brings it into the range at which *vegetative* bacteria are killed. One method of milk pasteurization involves holding milk at 62.8°C for 30 minutes. This is enough to kill any vegetative pathogens and eliminate sufficient spoilage organisms so that the food will keep fresh for up to three days under refrigeration.

Fig 3.6 Pasteurisation plant *Courtesy: The National Dairy Council*

However, as the temperature rises above 65°C less time is needed to kill vegetative organisms until at 100°C all will be destroyed in two minutes. Only *spores* resist destruction by boiling water. To be sure of killing these resistant bodies, it is necessary to reach 121°C, the temperature employed for commercial sterilization in the canning industry.

Cooling foods

When the temperature is reduced *below* 10°C, the growth of organisms gradually slows down. Most pathogens cease multiplying at about 5°C but there are exceptions to this rule. *Listeria monocytogenes*, a much publicised cause of concern, is one of these microbes. It grows slowly but appreciably down to temperatures as low as 0°C.

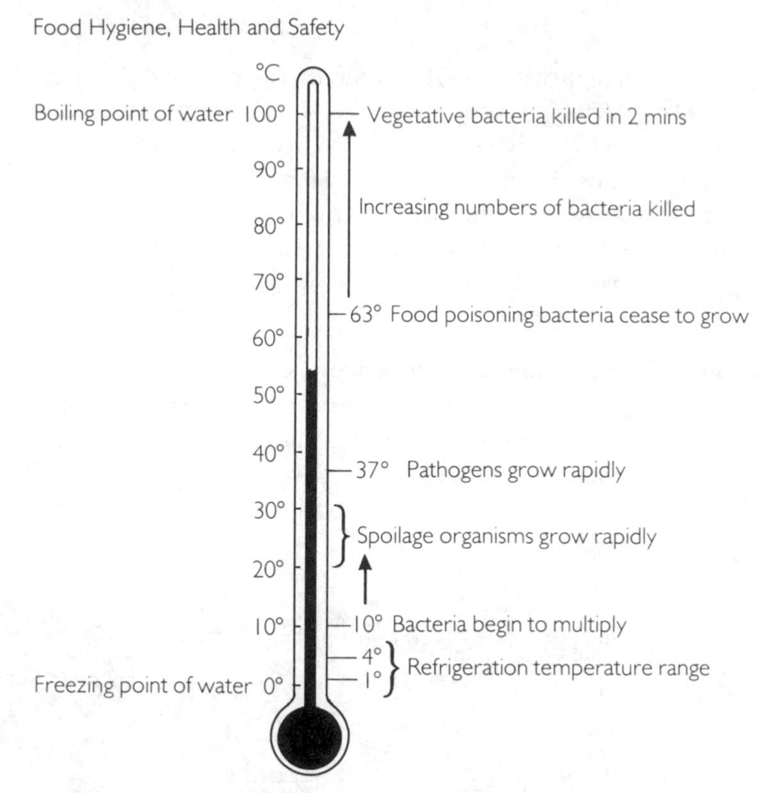

Fig 3.7 Effect of temperature on bacteria

It is important to realise that cooling to temperatures between 1° and 10°C only makes microbes **dormant** it does **not kill** them. When foods are removed from chill cabinets or refrigerators, they soon warm up. Once they reach 10°C, the previously inactive organisms in the food will begin to multiply once more. It is important to consider how long the journey from your supplier to the refrigerator actually is.

Note: Chilling is only a holding operation. ***It does not kill microbes.***

Chilling should only by used for *short term* storage of fresh foods. For long term storage, foods must be frozen to a temperature of –25°C or –30°C. It is important to understand that many organisms, particularly spores, *survive* even this drastic treatment and will *grow again* if the food is defrosted and left in warm conditions.

Oxygen

The majority of micro-organisms need oxygen to release energy from food materials such as sugars. Organisms which respire in this way are said to be *aerobes*. Certain types of cooking, processing and packaging remove oxygen from foods or replace it with other gases. This type of treatment discourages the growth of *aerobic organisms*. Examples of processes using this principle include:

- canning where all air is driven out in the process of sterilization – fish, meat, vegetables
- vacuum packing where the air is removed and the package sealed – smoked salmon, cheese, bacon
- packing in other gases as with sous-vide foods which are packed in a mixture of carbon dioxide and nitrogen.

Anaerobes A small group of micro-organisms respire *anaerobically*. They do not need oxygen and will, in fact, only grow in the *absence of oxygen*. Two species of food poisoning organisms belong to this group:

- *Clostridium perfringens* which causes symptoms of diarrhoea and vomiting.
- *Clostridium botulinum* which is the cause of a rare but deadly form of paralysis.

Both of these organisms produce spores, so we have to be aware of the danger of cooking methods like stewing, which drive out oxygen, but do not reach a high enough temperature to kill sporing bacteria.

In addition to the strict anaerobes there is a group of bacteria which are capable of growing in both aerobic and anaerobic conditions.

Effect of acidity/alkalinity on micro-organisms

The majority of food spoilage and all food poisoning bacteria thrive in near neutral conditions and are less likely to grow in acidic foods. This means that near neutral foods such as meat, fish, eggs and milk are the most vulnerable to spoilage by bacteria and are capable of supporting the growth of food poisoning bacteria. Moulds and yeasts on the other hand can tolerate quite acid conditions and so can spoil fruits, wines and some vegetables.

Note: Flesh, fowl and dairy products need hygienic handling and cool storage.

Neutral or low acid foods such as meats and vegetables can be made safe by pickling in vinegar, whilst high acid foods such as citrus fruits, rhubarb and gooseberries have a natural resistance to the growth of pathogens and spoilage organisms.

Task

Investigate the acidity/alkalinity of a variety of foods. This can be done quite simply by using indicator papers or, if available, using a pH meter.

Note: The pH scale measures the acidity or alkalinity of a solution. Those which have a value below 7 are acidic, those greater than 7 are alkaline and neutrality occurs at exactly 7.

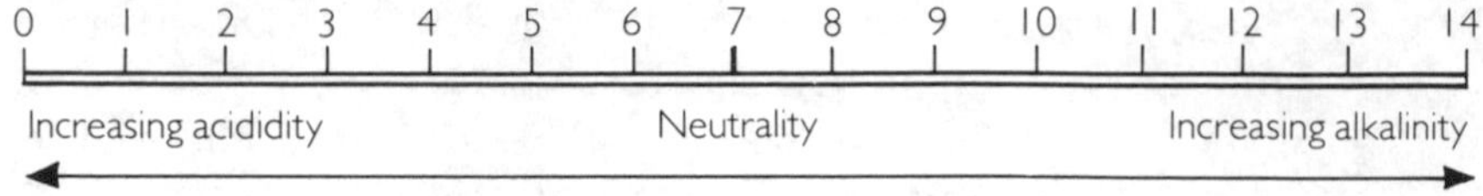

0 1 2 3 4 5 6 7 8 9 10 11 12 13 14

Increasing acidity Neutrality Increasing alkalinity

Fig 3.8 pH Scale

To make a food solution

Liquid foods like egg, milk and fruit juices can be used as they are. Just dip the indicator paper in the liquid and match the colour to the scale and note the pH number. Other foods must be ground up and dissolved in *distilled* or *deionized* water.

With foods like carrots, meat, etc. proceed as follows:

- Cut up the food finely, with a sharp knife.
- Grind chopped food in a mortar with some fine sand and a little water.
- Pour off the food solution. Filter through some muslin if necessary.

Test a variety of foods and then group their pH numbers as shown in Table 3.1.

You may also like to try foods such as bread, biscuits, pasta, and bicarbonate of soda.

Table 3.1

Group 1: Neutral foods (above pH 5.3)	*Group 2: Medium acid foods (pH 5.3 to 4.5)*
Meats	Turnips 5.2
Chicken 6.3	Carrots
Beef	Beetroot
Mutton	
Dairy products	*Group 3: Acid foods (pH 4.5 to 3.7)*
Butter 6.2	Bananas 4.6
Milk	Tomatoes
Cheese	Pears
Fish	
Salmon 6.2	*Group 4: High acid foods (pH 3.7 or less)*
White fish	Grapefruit 3.0
Vegetables	Oranges
Potatoes 5.4	Apples
Sprouts	
Cauliflower	

Time

Although bacteria and other micro-organisms multiply quickly, they need *time* to increase to numbers which will cause serious spoilage or food illnesses. If we take care to buy good quality foods, handle them hygienically, and store them at the correct temperatures, we will deny microbes the vital time to multiply to dangerous numbers.

Food poisoning bacteria vary, but a rough estimate of the number of bacteria needed to cause food poisoning is one million cells. Foods should never be as heavily contaminated as this if we follow three simple rules:

- Keep food clean – to keep the number of bacteria in the food to a minimum
- Keep food covered – to keep out dust and insects
- Keep food cold (below 10°C) *or* really hot (above 63°C) – to prevent the growth of bacteria

Summary of factors affecting growth of micro-organisms

Food
- Foods which are good for humans are good for microbes too!
- Preservatives can be used to inhibit growth of organisms in processed foods.
- Good hygiene standards reduce the numbers of microbes in fresh foods.

Water
- Micro-organisms need liquid water for growth and multiplication.
- Dried foods will not support microbial growth providing they are kept dry.
- The majority of microbes will not grow in high sugar foods.
- Freezing makes water unavailable to micro-organisms.

Temperature
- Microbes grow best in warm conditions.
- Cold foods are best kept below 4°C.
- Hot foods above 70°C.
- Vegetative bacteria are killed at 100°C for two minutes and above 65°C if longer periods of time are used.
- A temperature of 121°C is needed to be sure of killing spores.
- Chilling temperatures of –1°C to 8°C keep microbes dormant but do not kill them.
- Freezer temperatures (–25°C or –30°C) are needed for long term storage of fresh foods.

Oxygen
- Most microbes are aerobes i.e. need oxygen for respiration.
- Anaerobes do not use oxygen, and will not grow in the presence of it.
- A small group of bacteria are capable of growing in either aerobic or anaerobic conditions.

Acidity/alkalinity
- Most micro-organisms grow best in near neutral conditions.

Time
- Micro-organisms need time to multiply enough to cause food poisoning or serious spoilage in a food.

Task

After considering the factors necessary for growth of microbes you should be able to work out which foods need particularly careful handling because they can support the growth of food poisoning bacteria and which are regarded as 'safe' foods.

Table 3.2

Foods likely to support the growth of bacteria	Foods unlikely to support the growth of bacteria
Flesh foods	*Acid foods*
1 Beef	1 Lemons
2	2
3	3
Made up meat dishes, meat products	*Salted foods*
1 Meat pies	1 Anchovies
2	2
3	3
Meat juices, sauces	*High sugar foods*
1 Stocks	2 Syrups
2	2
3	3
Products containing fresh egg or milk	*Dry foods*
1 Cream	1 Biscuits
2	2
3	3

Place the following foods in the appropiate part of Table 3.2:
Jams, gooseberries, honey, sausages, pork, salted cod, gravies, pasta, custards, pickles, bacon, flour, milk, trifles, pasties, rice, and poultry.

Self assessment exercise

1 List the factors which affect the growth of micro-organisms.

2 Why do manufacturers add preservatives to some of their processed foods?

3 State three ways the water in a food can be made unavailable for the growth of microbes.

4 Why is it important to protect foods such as flour and milk powders from the moisture of the air?

5 If a microbe uses oxygen in respiration it is an ____________ . (Supply the missing word.)

6 Match the correct temperature or range to each of these five statements:
(a) Microbes do not multiply above this temperature.
(b) Minimum temperature to kill all spores.
(c) Optimum growth range for pathogens.
(d) Danger zone for multiplication of microbes.
(e) Refrigerator temperatures.

35°C-37°C   63°C   25°C-30°C   121°C   10°C-63°C   1°C-4°C

7 Explain the effect of chilling on micro-organisms.

8 From the list of foods provided, pick out two examples of each of the following categories:

(a) Neutral foods, likely to grow pathogens.

(b) Acidic foods, unlikely to grow pathogens.

Poultry cucumber pickle milk oranges bicarbonate of soda

9 Describe two ways you can prevent micro-organisms having the time to multiply to dangerous numbers in the foods you are preparing.

How food poisoning organisms get into the kitchen

As bacteria are invisible to the naked eye, we have to know where they are before we can work out safe routines for handling foods. Food poisoning bacteria can reach the kitchen in a number of ways and these are discussed in the following sections.

In the foods themselves

Some foods are likely to be infected *before* they reach the kitchen. These are the raw animal foods such as meat, poultry, fish and eggs. These foods can easily cause *cross contamination* that is, they can pass food poisoning bacteria on to cooked foods. Other foods which may bring infection into the kitchen are potatoes, root vegetables and mushrooms which may have *soil* clinging to them. Soil is the natural home of several types of food poisoning organisms, therefore, any vegetables which are not supplied prewashed should be cleaned well away from other foods and the soil flushed down the drain so it cannot contaminate other foods.

From people

We all carry thousands of bacteria on the surface of our skin, in the nose and throat, and internally in our intestines. Most of them are harmless but some can cause food poisoning if they get into foods.

Apparently *healthy* people can infect foods if:

- they touch foods with dirty hands – particularly if they do not wash their hands thoroughly after using the toilet.
- they sneeze, cough or spit near unprotected foods.

People who are recovering from food poisoning are very dangerous to have near food, especially if their personal hygiene is inadequate.

People can be *carriers* of food poisoning without being aware of their condition. Carriers continually excrete small numbers of bacteria in their faeces and so are very likely to infect the foods they handle.

From animals

Pet animals such as dogs and cats can carry pathogens in the same way as ourselves. Neither they, nor their food, should be allowed into any room where food is prepared, cooked or served.

Insects, rodents and birds

The food and warmth of catering premises attracts many unwanted creatures, particularly rodents and insects, and sometimes birds. All these pests can contaminate food with pathogens. Food premises need to be carefully built and maintained to keep out pests as far as possible. Good cleaning routines also help to keep infestation to a minimum. (*See* Chapter 13 for further information.)

Dust

Spores of pathogenic bacteria survive for long periods, getting a free ride on dust particles. Regular wet cleaning of floors and work surfaces, helps to cut down this type of contamination. Foods should be covered to prevent dust settling on them.

Equipment and work surfaces

It is essential that *everything* that comes into contact with food is clean. Since we know that food poisoning organisms may be present in many fresh foods, we must be sure that *separate* equipment is used for *raw* and *cooked* foods – otherwise cross contamination can take place.

After preparing meat, fish or eggs always ensure that all equipment and work surfaces are thoroughly washed in hot water and detergent before reuse.

Refuse and waste food

Bacteria multiply rapidly in refuse and waste food, making it a danger and source of contamination in the kitchen. Waste must be separated from fresh food and taken out of the kitchen, *regularly*. Staff who dispose of waste must wash their hands before resuming other work.

Cloths

All kinds of 'cloths' are used in kitchen hygiene. If we are not careful in their use, they can do more harm than good. Just imagine what happens when you use a swab to 'clean' a work surface. If you wipe it over and then leave the cloth sitting in a pool of water in the sink, in the warm, moist cloth, the bacteria will multiply very rapidly. When you pick up the cloth and wipe a table with it later in the day, are you *cleaning* the table or *putting more bacteria on it?*

Where possible use disposable cloths. If this is not possible, wring them out in hot detergent water after use and hang up to dry. Boil them at the end of each day.

Summary

Food poisoning bacteria get into the kitchen from:

- the foods themselves – especially raw foods
- the food handlers
- pests – rodents, insects, birds
- dust in the air
- cloths – towels, dish cloths, kitchen rubbers.

The food poisoning chain

Food poisoning does not occur as a result of one mistake or oversight in the preparation of a meal. It happens as a result of a *series* of events. The bacteria have to get into the food in the first place and they must also find suitable conditions in which they can then multiply.

The links in the food poisoning chain

1. Bacteria get into the kitchen.
2. The bacteria contaminate food.
3. The food is not thoroughly cooked.
4. The food is left in *warm* conditions for enough *time* for bacteria to multiply in the food.
5. The food is eaten.
6. The consumers suffer food poisoning symptoms between one and 48 hours after consumption.

How to break the food poisoning chain

1. Keep food premises as clean as possible – clean hands, clean equipment, clean work surfaces.
2. Prevent cross contamination – use separate equipment for raw and cooked foods.
3. Cook foods *thoroughly* but remember that spores can survive mild methods of cooking.
4. Serve piping hot as soon as possible after cooking, or cool quickly to 10°C and refrigerate or freeze.

To prevent food poisoning occurring we have to stop the transfer of pathogens to foods. Those foods which are to be eaten hot should be cooked to the correct temperature and then kept above 63°C until served to the customer. If this is not possible, the food must be cooled to 10°C *as quickly as possible* and kept refrigerated until needed. For prolonged storage, freezing is preferable for most foods.

Note: Good hygiene and control of temperature and time breaks the food poisoning chain!

The chain which leads to food poisoning	*Prevention of food poisoning*
1 Bacteria in the kitchen	1 Cut down the number of bacteria
2 Bacteria get into foods	2 Clean equipment
3 Food not cooked thoroughly	3 Cook thoroughly to kill most bacteria then either:
4 Food left in warm kitchen	4 Serve piping hot soon after cooking *or* Cool quickly and thoroughly
5 Food eaten	5 Food eaten

Fig 3.9 The food poisoning chain

Self assessment exercise

1 Which of the following types of foods are likely to be already infected with food poisoning bacteria before they reach the kitchen?
Poultry, biscuits, unwashed carrots, beef, jams, eggs, unwashed potatoes, canned peas.

2 State *two* ways food handlers can infect foods with pathogens.

3 Why are dogs and cats not allowed in kitchens?

4 Name *three* kinds of pests that can infest food premises.

5 How can you prevent dust contaminating foods?

6 What is meant by cross contamination? How can it be prevented?

7 Outline the procedures to be followed after cooking if the food is to be eaten cold.

8 Explain the dangers if these procedures are *not* followed.

4 Small equipment – design and safe use

Saucepans

Saucepans for use in commercial kitchens are designed for specific purposes i.e. for cooking various types of food by different methods. Their shapes and sizes vary from the small round ones with which you are familiar, to very large ones which appear to be of unusual design. The saucepans in everyday use in large kitchens are described here, but others, which are used less frequently, are not mentioned. The French terms are generally used to indicate which pan is required for a particular purpose.

Saucepan, Russe

These saucepans are made in various sizes from 500 ml-12 l. The larger capacity saucepans have an extra grip handle. This design is used for cooking vegetables, making sauces and stews and for boiling liquids.

Always use the saucepans with care and follow the basic rules for saucepan safety listed later in this chapter.

Fig 4.1 Correct hold for carrying or lifting a large saucepan when a loop handle is not fitted

To carry these saucepans if no grip handle is fitted:

1 Place your cloth down the length of the handle.
2 Place your arm along the top of the handle and grip the base of the handle.
3 Use the other hand to grip the base of the handle to give balance and extra support.

Sauteuse

A sauteuse is a shallow pan with sloping sides made in various sizes and has only one long handle. It is used for sweating or cooking food which may need to be tossed during working. The sloping sides assist the tossing movement.

Never overfill the sauteuse with oil or food otherwise the food will spill on the stove and the oil will splash or burn you.

Sauté pan (Plat à sauter)

Sauté pans are shallow with straight sides for shallow frying meats such as chicken, cuts of lambs and beef. The juices from the meat are retained in the pan and wine or sauce is added to give flavour to the meat sauce. These pans must not be used if food is to be tossed.

Never overfill the pan when cooking and avoid splashing when you place the food in the pan to cook.

Stew pans

Stew pans have two grip handles and short sided lids which come over the outside of the pan. The lids have two grip handles. The pans are used for making stews and braising meats. They may be used in the oven.

Never overfill – remember you may be using them in the oven. Always use *dry* oven gloves or *dry* cloths when removing from the oven. Indicate hot handles by sprinkling with a little flour.

Braising pans (Braisières)

These may be round, rectangular or oval with long grip handles and lids with a central grip handle. Used for braising or stewing, usually in the oven. Safety points are as for stew pans.

Stockpots

Stockpots range from 2.5 l–18 l and are fitted with two ring handles and a tap at the bottom. They are used for making bone stock and, sometimes, for cooking poultry.

Never overfill these pots because stocks are greasy and if allowed to boil over can become dangerous, as the fat can ignite as well as causing slippery spillage on the floor. The pots should always be used on stockpot stoves or low boiling tables to prevent unnecessary lifting.

When draining:

- avoid splashing
- avoid spillage
- avoid over-filling the container in which the stock is being drained.

Note: Stock must **never** be left in the kitchen overnight – It **must** be drained, cooled and refrigerated **as soon as possible**.

Saucepan safety

- Always place saucepans well on the stove, never on the edge.
- Always make sure the handles are turned inwards so that they do not cause obstruction.
- Always indicate hot handles by sprinkling with a little flour.
- Use a cloth to lift or move saucepans onto or off the stove.
- Never struggle to carry saucepans – always ask for assistance if they are too large or too heavy for you to lift.
- Never fill pans so full of cooking liquid so that when the food to be cooked is added the liquid flows over the top.
- Never use a pan which is too large or too small for the quantity of food or liquid.
- Never pull or push saucepans roughly on the stove – the contents may spill over onto the store, you or another food handler.

Cleaning saucepans

1 Remove all food particles, particularly from the corners.
2 Scrape and leave burnt saucepans to soak, with salt or soaking agent added to the water.
3 Wash thoroughly in hot detergent water paying attention to food particles in and around the rivets of the handles.
4 Rinse and sterilise. Leave to air dry and stack upside down on pot racks.

Cleaning copper

Copper utensils must be cleaned using a pickle of vinegar, salt and fine silver sand. Industrial rubber gloves must be used when applying this mixture. This process must be followed by thorough rinsing and sterilising.

Copper must never be allowed to become stained with *verdigris* – which is a green substance produced by the reaction of certain foods on the copper metal.

Copper must be relined when necessary.

Frying pans

These are made of black wrought iron and come in various shapes and sizes, designed for the specific purposes which the names indicate. The sloping sides allow foods to be tossed e.g. croutons, fried onions, pancakes and omelettes.

The oval meuniere pan is suitable for shallow frying fish fillets.

Frying pan safety

- Handles get hot – always use a cloth
- Never overfill with food or too much oil
- Never leave handles jutting out from the stove
- Never place any frying pan (or saucepan) with fat or liquid in it, in the pot rack above the stove. This can result in a serious accident if you take it down from the rack and hot liquid splashes over you or the hot stove.

Cleaning

1 Sprinkle with salt and heat gently.
2 Rub with kitchen paper or an old cloth to remove all food particles
3 Wipe out with a clean cloth and lightly oil before putting away.

Baking sheets

The wrought iron baking sheets used in pastry work should not be washed unnecessarily. They are cleaned by scraping with a metal scraper and wiped clean whilst warm with a thick cloth kept for this purpose. If they require washing, use hot detergent water, rinse and dry. Lightly oil to prevent rusting.

Tartlet moulds

These are made of tin plate and should not be washed but wiped with a clean cloth whilst warm. Only wash if absolutely necessary when they may require soaking before washing in hot detergent water, rinsing and drying. They are best dried in the oven after it has been switched off.

Sieves, strainers, graters and mashers

The following items require extra care in cleaning to ensure all food particles have been removed from the meshes and holes. Food must *never* be allowed to remain or accumulate on the meshes – an easy way to cross contaminate foods. If the food is allowed to dry or congeal it is very difficult to remove. *Wash immediately after use.*

Items difficult to clean

- sieves
- strainers
- graters
- mashers
- whisks
- piping bags

Cleaning

Use hot detergent water, and a short bristled brush designed for this purpose, to scrub the food from the holes. Run under hot water to rinse, and dry thoroughly.

Avoid soaking sieves which have wooden frames as this causes the wood to warp, split and harbour bacteria.

Small metal tools

These include whisks, spoons, ladles, fish slicers, skimmers, wire spiders, food tongs and scoops. They are made of stainless steel or tinned steel and require special care when washed. Pay attention to the handle where the prongs meet and where the prongs cross, as food is easily caught and soon congeals in these areas.

Wash in hot detergent water and dry thoroughly. If moisture is allowed to remain in the whisks, the next time they are used globules of dirty water splash into the foods. The other items, such as skimmers, are easier to clean but pay attention to the perforations to ensure that no food particles remain. With tinned steel items once the tinning has worn through they must be retinned or discarded. Hang these items by the hook handles on hooks in the equipment store.

Wooden utensils

These include wooden spoons, spatulas, chopping boards, mushrooms, cheeseboards. Wooden items are made of hard wood and are constructed in one piece.

Chopping boards are used for light chopping e.g. vegetables, salad items, parsley. *Never* use them for chopping meat or poultry because the chopping action will cause splinters which find their way into foods. Food particles collect in the cuts and bacteria multiply. To clean – scrape away any food particles, wash in hot detergent water, rinse and dry. Stack them in such a way that air can circulate and dry them completely – **never** stack them on top of each other.

Spoons, spatulas and mushrooms are washed in hot detergent water, rinsed and dried thoroughly before putting away.

Note: Destroy any wooden items which are split and showing signs of wear.

Fish scissors, larding needles, trussing needles

These items are used in the preparation of raw fish, meat and poultry so need care in cleaning to prevent contamination. Some types of fish scissors are designed so that the two blades can be taken apart for cleaning. They should be washed in hot detergent water, dried and then reassembled. Take care to remove the fish scales – they dry and stick to knives and scissors.

Larding needles are used to sew strips of raw fat into meat so it is very important to keep them clean and hygienic. The needles are hollow and the fat is placed inside. An implement is provided with each set of needles for cleaning out the hollow. After removing any particles of fat or meat, wash thoroughly in hot detergent water, rinse, shake any water from the hollows, and then dry. Sterilise if necessary.

Trussing needles are like enormous sewing needles and are used for tying or trussing poultry. Wash thoroughly in hot detergent water, rinse and dry.

Note: The points of larding and trussing needles are very sharp. They should be protected by a cork when not in use.

Butchers' cleavers, butchers' saws

Cleavers

These are designed for chopping through small bones, mainly lamb and pork. They are available in various weights.

Safety points

- Stand comfortably in front of the joint
- Choose a cleaver which is the right size for you to use safely – neither too heavy nor too large.
- Make sure your hands are free of fat and moisture so you have a firm grip.
- *Never rush* – take your time.
- Use an even chopping motion with the tip of the cleaver doing the cutting.
- Keep your eye on the cut and your *hand well clear.*
- Rest when your arm gets tired.
- Always chop on a butcher's block designed for the purpose.

Washing Remove meat particles from the cleavers. Wash in very hot detergent water, rinse and dry. Sterilise if necessary. Store in a drawer or hang them safely where they can be seen easily.

Saws

Bow saws are used for cutting bones which are well covered by meat and tenon saws for bones which are on the surface.

Safety points

- Stand comfortably in front of the meat.
- Check that your hands are free of grease or moisture.
- Draw the saw backwards and forwards. *Do not force the saw*, if you do it will jump and cut your hand.

Washing Remove all meat particles from the teeth of the blades. Wash thoroughly in hot detergent water, rinse and dry. Sterilise if necessary. Store as for cleavers.

Tin openers

The hygiene of tin openers, particularly the table models, is badly neglected in most kitchens. The table models are used for opening a wide variety of tinned goods such as fruits, vegetables, meats, and fish in oil and sauces. All these foods come into contact with a blade which is rarely washed. In addition, fragments of metal may adhere to the blades and contaminate foods.

The opener should be throughly washed after *each use*. If this is not practicable, it should be washed at least twice a day – more often if it is used very frequently or for a variety of foods. Wash with detergent water, rinse and dry. Pay extra attention to the blade and the rotating cog. After each cleaning, place a little food machine oil on all the moving parts.

Piping bags

These are made of calico, plastic or nylon and are use for piping cream, potato mixtures, raw fish and chicken mixtures. You can see immediately the dangers of cross contamination which can arise if piping bags are not cleaned thoroughly after each use.

How to clean piping bags Nylon bags:

- Squeeze out the contents, turn inside out and remove the piping tube.
- Wash away all food particles and rinse thoroughly.
- Sterilise, using a suitable sterilant e.g. Milton.
- Allow to dry thoroughly before use or putting away.

This method can be used for plastic and nylon bags.

Cloth bags:

- Follow steps one and two as for nylon bags.
- Boil the bags with a suitable detergent for 5-10 minutes.
- Rinse and dry the bags thoroughly before use.

Do not boil nylon bags.

Cloth bags are more hard wearing than nylon ones but require more thorough cleaning. Nylon bags discolour easily.

*Note: Piping bags must be washed **thoroughly** after each use. **Never** use a dirty piping bag.*

Piping bags should be marked or colour coded according to their use e.g. fresh cream, fish, forcemeat (stuffing), poultry or veal forcement. This should prevent mistakes such as using a bag for fresh cream which has been used for forcemeat.

Disposable piping bags are now available. They are more hygienic if used properly and cut out a lot of work.

Notes

There are many other items of kitchen equipment too numerous to mention here but if you understand the methods of cleaning and care of the items above, you should be able to deal with the other types of equipment.

Always *think* before you use a piece of equipment. It is dangerous to use any equipment for a purpose for which it was not intended or designed – **never** do it.

Electrical equipment

Electrical equipment is commonplace in the modern kitchen. It makes your job easier but, at the same time, great care and concentration are required whilst operating, dismantling, cleaning and reassembling this type of machinery.

The items of electrical equipment you will find in most kitchens are, food processors, food mixers, mincers, vertical cutter mixers, food slicers, bowl choppers and chipper dicers.

General rules for operating

Ensure that:

- the machine is suitable to carry out the task you want to do.
- the machine is assembled properly and is in working order.
- all guards are in position and firmly connected.

- that the machine has been properly cleaned before you use it.
- that any safety locks have been operated.
- You know how to use the machine safely – if not ask for assistance
- you are correctly dressed – no earrings, bracelets, bangles, loose apron strings or a neckerchief that could get caught in the machine.
- the machine is in neutral or that you have selected the correct speed or thickness of cut you require.
- you have the equipment required to place the food on or into the machine
- it is safe for you to commence work – there are no surrounding hazards which could affect your operating the machine safely.
- you have selected the correct blade or cutter or attachment to do the job properly and safely.

Cleaning procedures

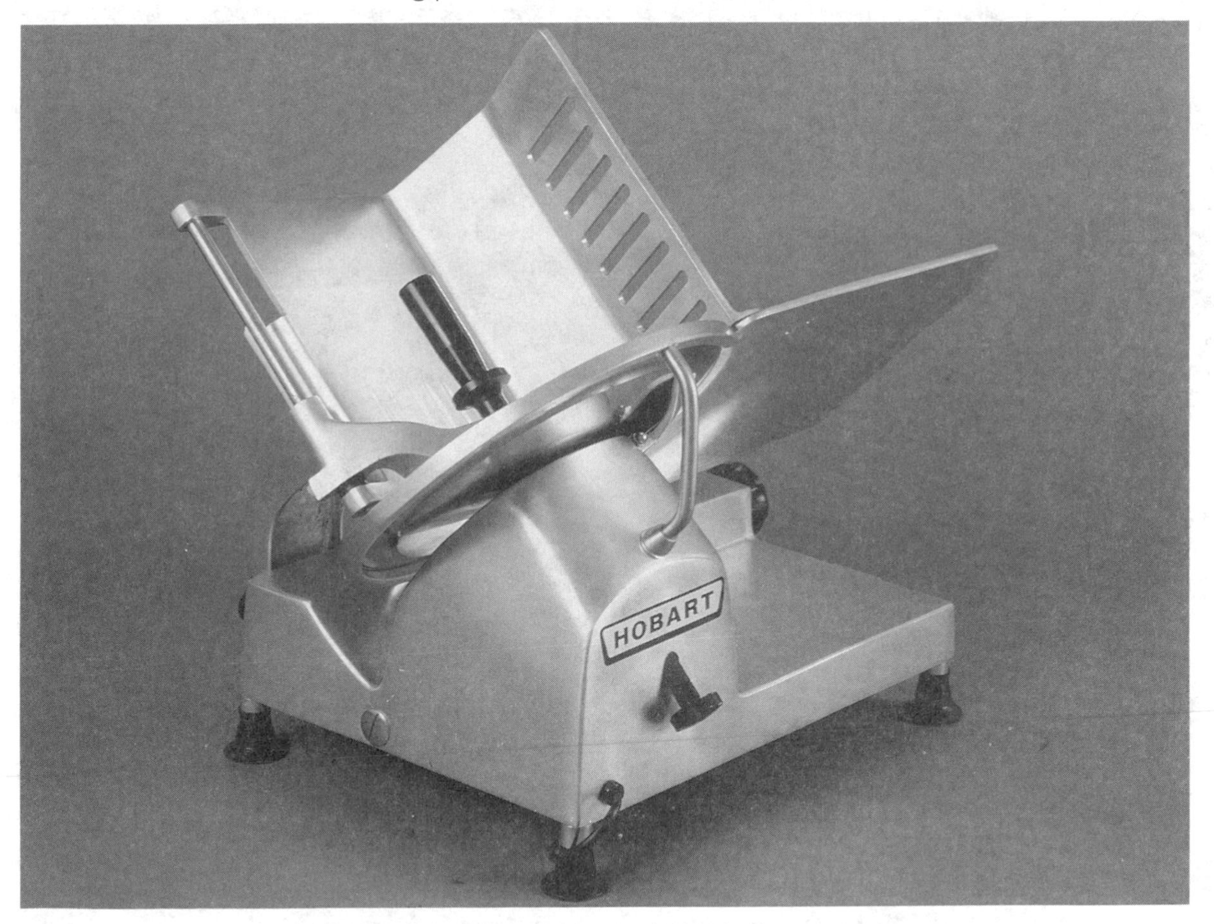

Fig 4.2 Food slicer requiring care in cleaning

1 Switch off the power at the mains and unplug or use the isolator switch to disconnect.

2 Dismantle the machine according to the manufacturer's instructions or in the sequence you have been shown.
3 Handle all moving parts with care, particularly blades, cutting discs and shredders.
4 Remove all food particles from attachments.
5 Wash in very hot detergent water, rinse and dry thoroughly. Use a sanitiser for cleaning the fixed blades of slicing machines.
6 Pay particular attention to corners and crevices where food is liable to accumulate and ensure that all traces have been removed. (*See* Fig 4.2.)
7 Reassemble the machine in the correct sequence, checking that each piece is clean. Secure all guards and safety catches.
8 Check the machine is in correct working order before you leave it.
9 Switch off at the mains.

Fig 4.3 Machine attachments requiring care in cleaning

Safety of electrical equipment

- Always clean immediately after use and before processing another food.
- If unsure of the correct procedure of cleaning or dismantling, always ask your supervisor to check your work until you feel you have mastered the skills required for the equipment you are using.
- All machines should be purchased with a regular servicing contract with the manufacturer or servicing agent.
- The servicing of machines should be entered in a book kept for the purpose. The details which should be entered are:
  1 type of machine and number
  2 details of faults
  3 servicing carried out
  4 date
  5 signature of servicing engineer.
- Report any defects of the machines *immediately* to your supervisor who will put the machine out of use until the fault has been rectified.
- All attachments must be stored safely and correctly e.g. cutting blades and discs and appropriate tools should be hung on brackets or hooks on the wall by the machine. The parts can be seen easily and any missing pieces spotted immediately.
- Only use the attachments for the specific purpose for which they were designed. *Never* improvise without thinking it through first. For example, a knife handle used as a plunger in a mincer could easily cause the blade to be pulled from the hand — resulting in injury.

Microwave ovens

Microwave ovens are designed for both the domestic market and industrial use. The domestic models are not suitable for heavy commercial use so must not be used in catering establishments. Professional advice should be sought from manufacturers as to the suitablity of these ovens for particular catering operations before purchase and installation.

Use

- Familiarise yourself with the controls and operation of the oven.
- Always follow the manufacturers' instructions, particularly regarding:
  1 the use of turntables and racks which ensure that the food is evenly heated.
  2 the standing time necessary for the operation. Standing time is the extra time required after cooking, to spread the heat evenly throughout the food.
- Do not overload the oven with food as this will result in uneven cooking or cold spots in reheated foods.

- Microwave ovens, if used correctly, will cook foods and regenerate ready-cooked foods which have been chilled or frozen. They are *not* designed to *sterilise* foods.
- Regenerated or cooked meat dishes must reach a minimum temperature of 70°C throughout – with no cold spots. Some microwave ovens have built-in probe thermometers which can be used to check that the dishes reach the required temperature. If this facility is not provided a specially designed microwave thermometer can be inserted into the food.
- Regular servicing by qualified engineers is essential to ensure that the ovens continue to work safely and efficiently.

Lighting gas stoves and ovens

Most modern ovens and solid top stoves have pilot ignition with a fail safe device. This ensures that the pilot is lighted before the main burner. Should the pilot light fail, the gas to the main burner will cut off preventing it from escaping into the kitchen, and perhaps causing an explosion. The ignition systems on the different makes of stoves vary, but the principle is the same.

General rules for lighting stoves with pilot light ignition

1. Remove both rings from the centre of the stove.
2. Turn on the gas supply to the pilot, press the ignition button and apply a lighter to the nozzle. When the gas is lit, keep the button pressed in for a further 30 seconds.
3. When the pilot is alight, turn on the gas to the main burner.
4. Replace the outer ring but leave the inner ring off to ensure the burner has ignited properly and any unlit gas has dispersed.
5. Replace the inner ring and adjust heat as required.

Lighting gas ovens

1. Light the pilot as explained for stoves.
2. Turn on the main gas supply to the fullest extent, usually Mark 9. This should light the oven.
3. Check, using the inspection hole, that it is lit properly.

Some models have electric ignition switches which create a spark to light the pilot.

***Note:** when you open oven doors, be careful of heat blast.*

Open top stoves

Open top stoves and boiling tables do not have pilot ignition. To light them, turn on the gas supply and apply a lighted taper to the inner ring and then to the outer ring. Adjust as necessary.

Steamers

Some steamers create their own steam by heating water in an internal chamber. Others have an external supply of live steam. Most gas models have pilot ignition.

When lighting steamers ensure that the gas has lit properly, that the water chamber is full and that the inlet valve is in working order. The chamber will be empty if the valve is blocked, for example.

When opening the steamer, release the pressure *before* opening the door. Make sure nobody is standing near the steam outlet.

Fat fryers (fritures)

Modern fryers have pilot ignition and are thermostatically controlled to prevent the fat/oil overheating.

When filling a fryer, make sure that it is filled only to the recommended level.

Draining fritures and straining oil safely

The oil has to be drained from a friture after use, to allow the oil to be filtered and the fryer to be cleaned. Some safety points to note are:

- Allow the oil to cool before draining.
- Remember the container for the oil will be hot.
- Make sure you have closed the drain tap before refilling.
- Take care when tipping the oil back into the friture. If you spill any oil on the floor, make sure people are *aware* of the hazard and clear it up *immediately*.

Opening cans

Food cans are made in many different shapes and sizes to pack a wide range of ready to use foods. Some tins of ham and various types of fish are supplied with keys.

To open an oval can (such as a tin of anchovies) with a key safely:

- Place the cut away or grooved side of the key on the tag.
- Make sure the key is placed on the correct side of the tin so it is free to turn.
- Make sure you have a good grip on the can and take care not to spill the liquid.

Tins of ham, even if fitted with a key, are best opened with a wheel-type of can opener. Open both ends and push the ham through the can.

General rules for opening cans

- Make sure the can is properly supported either by the magnet attached to the opener or on a table.
- Always remove the lid of the can *completely* by cutting all the way round.
- If in spite of your efforts the opener has not removed the lid cleanly, use a cloth and remove the lid manually by bending it from side to side.
- **Never** leave a tin with the lid still attached and raised above the rim.
- If a can is *blown* i.e. has bulging ends, *do not* attempt to open it. The contents have deteriorated and could be poisonous (*see* p 88). If you were to puncture the can the contents would spurt out over you and might also contaminate any food nearby. Blown cans should be returned to the supplier who will credit the cost to your company.
- Slightly dented cans are safe to use *providing* there is no indication of damage or puncturing. Cans which show rusting or seepage should definitely *not* be used.
- Canned foods, once opened, must be used immediately or transferred to a basin for storage in a refrigerator. Food must never be left in an open can for storage.
- All food cans must be rinsed after being emptied to prevent attracting flies and other pests when discarded.
- **Always** wash the tin opener after use.

Hazard spotting

This chapter has covered many aspects of safety in the kitchen, but information alone will not prevent accidents. You need to develop an awareness of the *causes* of accidents so that you can prevent them happening. If you are alert and observant you can spot potential hazards and take action *before* accidents occur.

Exercise

Make a checklist of potential hazards in your working environment and check them on a regular basis. A suggested checklist is given below. Try it out in your workplace. How does it rate? Where are improvements needed?

SAFETY IN THE WORKPLACE

Machine operation

Are machines being operated safely? Are the guards used?

Are staff suitably dressed?

Power points

Are several appliances working off one point? If so, has the total wattage of the appliances connected to the single point been checked for safety?

Are the adaptors in good order?

Working practices

Are incorrect knives or utensils in use? Are blunt knives in use?

Is the working area untidy?

Floors

Is there spillage of fat around stoves or fryers?

Are the floor surfaces poorly maintained? Are any floor tiles loose?

Are gangways and passages clear and if not, what is obstructing them?

Firedoors

Are firedoors wedged open?

Are they locked or obstructed?

Waste bins

Are they filled to overflowing so they are heavy to move?

Do they have lids on?

WORKPLACE HYGIENE

Ventilation equipment

Are dust and grease accumulating in the extractor fans and ventilation ducts?

Hand washing equipment

Are there nail brushes at each wash hand basin? Is there soap in each dispenser?

Refrigeration equipment

Are cold rooms and refrigerators clean or are there food scraps in the corners?

Are any of the working areas dirty? Is any of the equipment dirty? Is there an accumulation of grease around equipment?

Are staff incorrectly dressed – no hats or wearing outside clothes for example?

FOOD HYGIENE

Temperature control

Are the internal temperatures of refrigerators and cold rooms checked frequently to ensure the correct temperatures are maintained (1°-4°C)?

Are the hot cupboards and bain maries used for keeping food hot working at the correct temperatures or are they working below 65°C?

Are any foods such as cooked meats and poultry, soups, stocks, stews and sauces, left to cool in the warm kitchen?

Are any cold foods left unrefrigerated?

Are any high risk foods such as custard, cream, soups or stews, frozen at too high a temperature i.e. above –20°C?

Cross contamination

Is equipment used for preparing raw or cooked foods not cleaned properly? (Chopping boards, knives, ladles, whisks.)

Are short cuts taken under pressure of work in the kitchen?

Are these items clearly marked for use with raw or cooked food?

For instance, is a grapefruit cut with the first knife to hand without considering whether it has been previously used on raw meat or shellfish?

Is defrosting left to the last minute and done by unsuitable methods or in unsuitable places – e.g. by placing poultry in hot water, in a warm oven or on the rack above the stove?

Many of the hazards mentioned above contravene the Food Hygiene Regulations 1989 or the Health and Safety at Work Act 1984. See if you can identify the points from each Act.

The list above is only meant to start you thinking about kitchen hazards. Each kitchen is different so you will need to tailor your own list to your particular workplace. Leave space against each item to note the action you propose to take. Some items you will be able to deal with yourself, others will need the co-operation of your supervisor or employer. Remember to date your checklists so that you can see clearly when hazards have occurred, when remedied, and note those which still need attention.

Accidents do not just happen, they are caused. If the causes can be eliminated, catering will be a safer and healthier industry.

5 Food poisoning organisms

Food poisoning

Any illness which follows soon after the consumption of food can be regarded as 'food poisoning'. The illness is usually characterised by vomiting and/or diarrhoea as the body seeks to remove the poisons and prevent damage to vital organs such as the heart, lungs, liver and kidneys. The majority of cases showing these symptoms are due to the presence of bacteria, viruses, or their products in the food consumed. However, there are many other poisonous substances which may occasionally contaminate food and these may induce similar or related symptoms. Examples of these substances are given below.

Naturally poisonous foods

Many plants and some animals contain substances which are toxic to man. Early men had to learn by trial and error which natural foodstuffs could be eaten safely and which were poisonous. Some familiar examples of plants which are poisonous are:

The potato Tubers which have been exposed to the light become green and develop sufficient *solanin* to cause illness, or even death if eaten in large quantities.

Rhubarb The leaves contain *oxalic acid*, hence the need to remove the leaves and about one inch of the leaf stem when preparing rhubarb for stewing.

Red beans The beans contain a substance which attacks the red cells of the blood. The toxin is destroyed if the beans are fast boiled and the water discarded.

Fungi Fungi are another group of plants where careful identification is essential. In the past when mushrooms were picked in the wild instead of being supplied from specialist farms, the Death Cap fungus, *Amanita phalloides*, was sometimes confused with the common mushroom with invariably fatal results.

Fish and shellfish Some fish and shellfish can become very dangerous in certain circumstances:

- Fish of the mackerel family produce a histamine-like substance if they are stored without refrigeration or preservation. The substance causes a rash, vomiting, dizziness and difficulty in breathing in anyone consuming the fish. The disease is known as *scombrotoxic fish poisoning*.

- Oysters and mussels which have gorged on minute sea creatures called *dinoflagellates*, cause *shellfish paralysis*. The toxin paralyses the heart and respiratory muscles.

Food allergy

Some people react quite violently to foods which are harmless to the majority of the population. They are allergic to some chemical substance in the food concerned. The substance (allergen) causes the body of the affected person to produce antibodies which induce the symptoms of the disease – rashes, vomiting and other reactions.

Food contamination

Most of our food passes through a long production chain before it reaches our plates, so there are many stages where accidental contamination may occur.

At the farm

Farmers use a wide variety of herbicides, fungicides and pesticides to protect their crops and increase their yields. There are strict regulations which control the nature of these substances, the quantity used and the time they can be applied. Nevertheless residues of these substances may be found sometimes, at least on the outer surfaces of food crops.

Most crops are harvested mechanically so all sorts of unwanted materials may be mixed with food materials when they reach the factory – inedible parts of the plants, insects, soil, stones, even pieces of string or loose nuts and bolts.

At the factory

Food crops go through a number of cleaning processes to remove contaminants:

- **Screening** They may be passed through screens of different sizes to remove solid material.
- **Metals** Metal detectors are used to search for non-magnetic metals and electromagnets to remove magnetic metals.
- **Washing** Finally most crops are washed either by jets of water or by propelling them through a series of tanks by rotating paddles. Having obtained a clean food, constant care is needed to keep it free from accidental contamination during processing or packaging.

Contamination in the home or catering establishment

Foods can be contaminated by the water used for preparation or cooking, or by the containers in which they are stored or cooked. This is true of contamination from poisonous metals such as zinc, copper and antimony.

At the beginning of this century, most of the water pipes of domestic households and commercial businesses in the UK used to be made of lead. In soft water districts, lead dissolved in the water. Lead is dangerous because it is not eliminated from the body and can accumulate to dangerous quantities. This particular problem has been overcome by artificially hardening the water where lead piping is still in use, and by the use of less poisonous metals such as copper or plastic piping which has been superseding this in modern buildings.

Zinc is another poisonous metal. Galvanised containers contain the metal so should not be used for storing food or drink, particularly those which are acid and liable to react with the metal.

In the same way fizzy drinks can dissolve dangerous amounts of copper from drinks machines if the pipes are not regularly flushed with fresh water. Similar problems have been reported with ice-lollies, which picked up dangerous amounts of copper from copper moulds which had lost their protective tinning. Another poisonous metal, antimony, used in cheap enamelled ware has contaminated acids foods when they have been stored in chipped pans.

The late 1980s saw a growing concern about the use of aluminium ware for cooking food. The anxiety arose from the discovery that the abnormal cells in the brains of patients suffering from Alzheimer's disease have been shown to contain an excessive amount of aluminium. Alzheimer's disease is the most common form of senile dementia. However, it must be realised that aluminium is found widely in soils, so occurs naturally in many foods.

The Ministry of Agriculture estimates that the average intake of aluminium is about 6mg per day, of which 90 per cent comes from food and the rest from water. The amount of the metal picked up from the older (non-coated) type of aluminium saucepans is thought to be about 0.1mg per 100g in the case of most foods and 0.7mg from acid foods such as stewed apples. The amount of the metal taken in from non-coated aluminium pans is, therefore, very small compared to the aluminium already in the foods themselves. With the newer, coated type of pan it is claimed no aluminium passes from the pans to the food.

Researchers are only beginning to understand the way in which aluminium passes into the brains of Alzheimer's patients so it seems a sensible precaution to purchase coated pans when buying new aluminium ware, and to avoid using uncoated pans for cooking acid foods such as fruits.

There are obviously a large number of potential sources of chemical poisoning both from natural and industrial causes but the number of cases of this type is small because producers and manufacturers are aware of the dangers and take care to control the quality of their products. The majority of food poisoning cases are due to living organisms. Details of the main food poisoning organisms are given in this chapter and those which cause food-borne diseases are described in Chapter 8.

Salmonella

Bacteria belonging to this group cause two distinct types of disease in man:

1 The enteric fevers

Typhoid and paratyphoid are due to infection with *Salmonella typhi* and *Salmonella paratyphi*. The infections are usually associated with infected water supplies, but food may be the source as in the Aberdeen epidemic in the 1960s when meat was contaminated by infected cooling water at the cannery. These two diseases are described in Chapter 8.

2 Salmonella food poisoning

The remainder of the 1500 strains of *Salmonella* cause 'food poisoning' of the *infective* type. This means that the bacteria multiply in the food so that large numbers of *living* bacteria are present in the food when it is eaten. Some of the bacteria are killed by the acid in the stomach, but most pass through into the intestines where conditions are favourable for their growth. They multiply and eventually die, releasing poisons which cause the unpleasant symptoms of the disease.

Salmonella causes the most common form of food poisoning in Britain and the number of cases has been rising steadily.

Salmonella bacteria

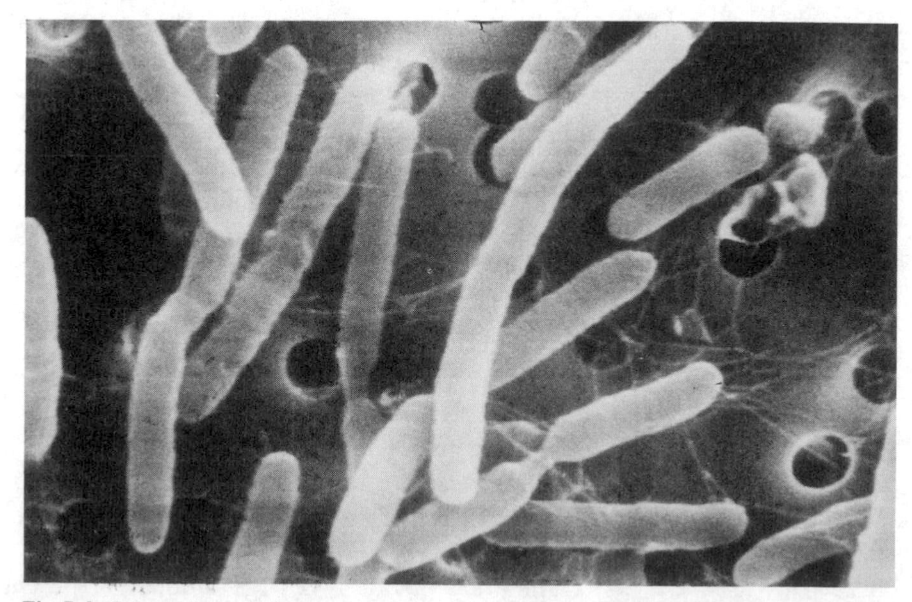

Fig 5.1 *Salmonella* bacteria *Courtesy: London School of Hygiene & Tropical Medicine/Science Photo Library*

Salmonellae are small rod shaped bacteria. They have flagella, small whip like projections, which propel them so they are said to be *motile*. They do not produce spores so are easily killed by normal cooking methods. Even exceptionally heat resistant strains of *Salmonella* are killed by holding a food at 70°C for 15 minutes.

The disease

It usually takes 12–24 hours after the consumption of the infected food for the symptoms of the disease to appear. This period is known as the *incubation period*. The time varies with the strain of *Salmonella* and how many bacteria are present in the food, so the incubation period can be as short as six hours or as long as 36 hours.

Diarrhoea is the main symptom of the disease but the patient is usually feverish and suffers from abdominal pain and vomiting. Symptoms last between one and seven days, sometimes longer. Occasionally some strains of *Salmonella* (e.g. *S. virchow*) get into the patient's bloodstream and cause a more serious and longer lasting form of the disease.

In most patients recovering from the disease, the number of Salmonella in the faeces drop, until finally they are entirely free of the bacteria. Unfortunately this is not so in all cases. About five per cent of people who suffer from *Salmonella* food poisoning become symptomless carriers of the disease. This means that they continue to excrete small numbers of the bacteria in their faeces – even when they feel completely well. It is obviously essential to find any individuals so affected in this way and exclude them from food handling while they are in this condition.

Sources of infection

Food handlers

People recovering from *Salmonella* food poisoning or carrying the bacteria can contaminate food if they do not wash their hands thoroughly after using the toilet.

Farm animals

Cows, sheep and pigs suffer from *Salmonella* infections so their meat and meat products, such as pies and sausages, frequently carry the organism. *Salmonella* has been found in unpasteurised milk and cream, and artificial cream has also been a source of this infection.

Duck eggs have long been recognised as a likely vehicle for *Salmonella* so it has always been recommended to use them only when hard boiled or in dishes where they would be thoroughly cooked. Until the late 1980s it was assumed that any *Salmonella* infection of hens' eggs would be only on the *outside* of the shell. It was, therefore, supposed that if a hen's egg was broken hygienically, the inside would be free from these bacteria. Unfortunately this is no longer a safe assumption since a strain of *Salmonella* has arisen (*S. enteritidis, phage type 4*) which infects the egg-laying apparatus of the hen. This means some hens' eggs may be infected with *Salmonella* **inside** the shell. The poultry industry is taking strenuous steps to eliminate this particular strain but it is still a source of infection which should not be ignored. (*See* eggs p 97.)

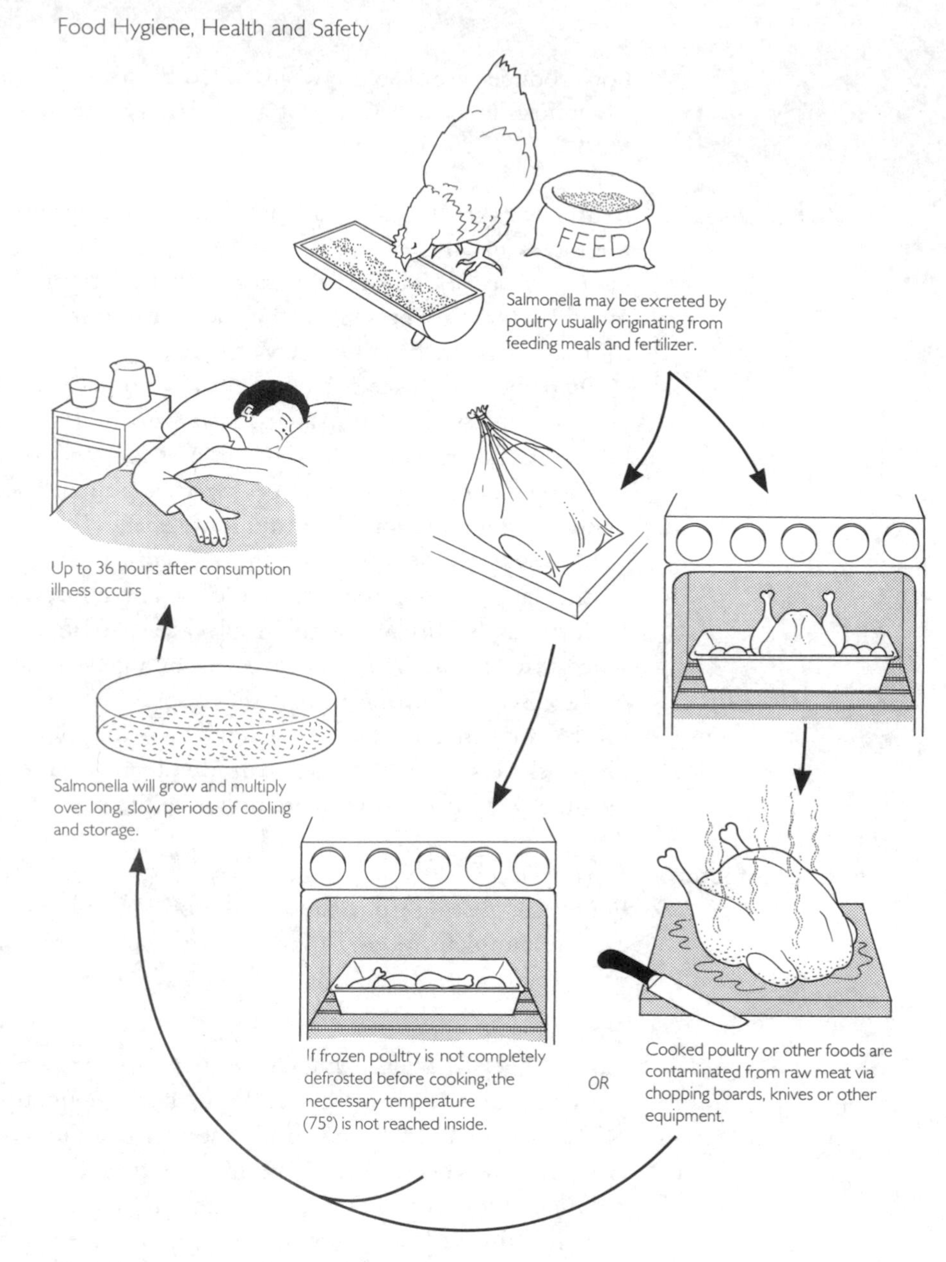

Fig 5.2 Sources of *salmonella* infection

Domestic pets

Dogs and cats and other domestic pets can be infected with *Salmonella* which is one reason why they and their food are banned from food rooms.

Pests

Rodents and insects such as flies and cockroaches can carry the infection, so food stores must be protected from these pests.

Prevention 1 Food handlers should be careful to wash their hands thoroughly (*see* p 13) especially:

- after visiting the toilet
- after handling raw foods such as meat, fish, poultry and eggs
- before handling foods such as cold sweets, cooked meats and salads which are not going to be cooked before consumption
- after handling swill.

2 All equipment – knives, cutting equipment and cutting boards – must be thoroughly cleaned and sterilised after use with raw foods. Preferably, separate equipment should be used to make cross contamination impossible.

3 Cooking and cooling food:

- Poultry and joints of meat must be completely thawed before cooking. (*See* p 115.)
- Stuffing must be cooked separately from the bird.
- Poultry must be cooked until every part of the bird reaches at least 70°C (158) preferably 75°C. (*See* p 105.)
- Serve hot food as soon as possible after cooking, ensuring that it stays above 63°C until it is consumed.
- If needed cold, cool *rapidly* to 10°C and keep refrigerated or frozen until needed. (*See* p 105.)

Clostridium perfringens

Clostridium perfringens causes the second most common form of food poisoning in Britain. The disease is usually mild though there have been occasional fatalities.

The disease caused by *Clostridium perfringens* shows some of the signs of both infective and toxic food poisoning. No toxin is produced whilst the bacteria are multiplying in the food but when the infected meal reaches the intestine, conditions are right for the production of spores. As the spores form, toxin is released causing the symptoms to appear.

The bacterium The bacterium is a rod-shaped sporing organism. It is anaerobic so grows readily when oxygen is at a low level or absent.

The disease The symptoms are abdominal pain and diarrhoea. Vomiting is rare and there is no fever. The symptoms usually last 12–24 hours. The incubation period is 8–22 hours.

Sources of infection The bacteria are widespread in the soil, dust and water and are carried in the intestines of man and animals. Spices are often heavily contaminated by this organism.

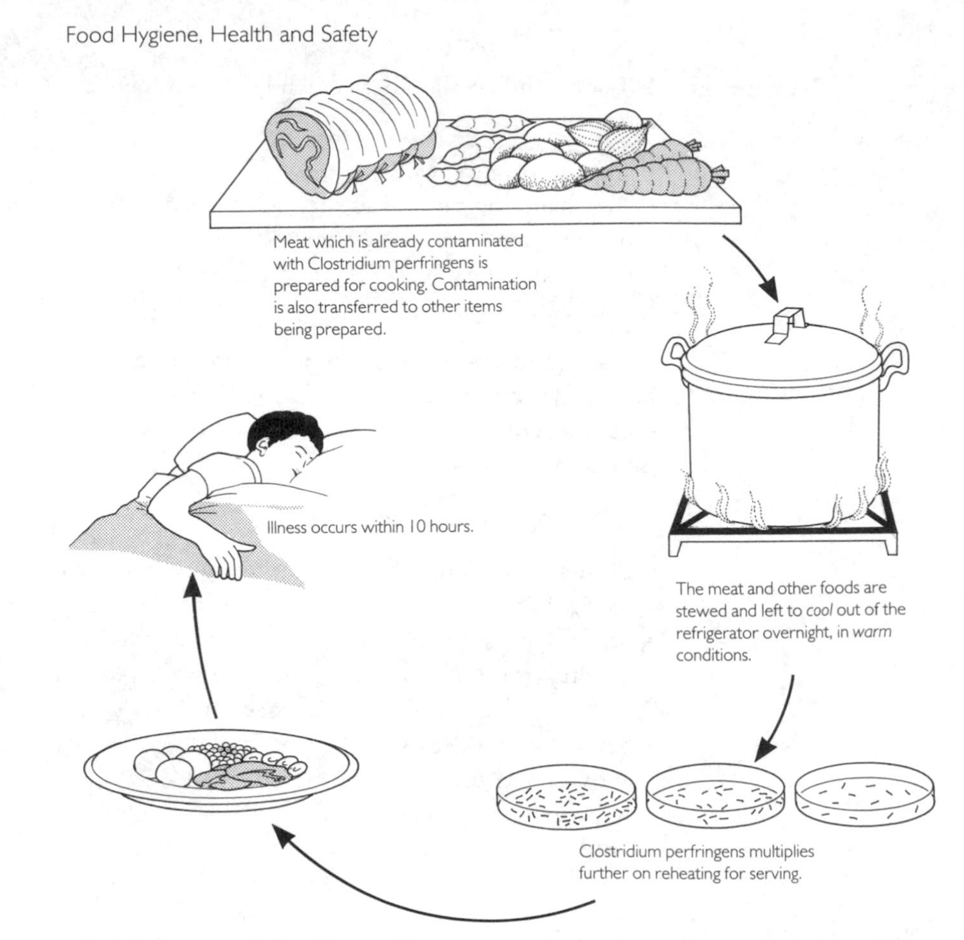

Fig 5.3 Sources of *Clostridum perfringens*

It can come into the kitchen:

- in meats and meat products
- in the soil clinging to root vegetables
- in dust which can settle on foods left exposed to the air
- on the hands of kitchen staff, if their personal hygiene is poor.

Types of food associated with the disease

This type of food poisoning generally occurs when large quantities of meat dishes are prepared for consumption on a subsequent day. The cooking has usually been done by low temperature methods such as steaming, stewing or braising. Often large joints have been cooked in this way. The centre of a large joint of meat is ideal for the multiplication of *C. perfringens*. The oxygen content is low and heat penetrates to the centre of a large joint only very slowly. Stews and casseroles are also ideal media for the growth of this organism as cooking drives off oxygen from the stock the meat is cooked in. Thorough cooking will eventually kill the vegetative bacteria but not the spores as they can stand boiling for several hours. The spores remain alive, though dormant.

If the meat is served soon after cooking, no harm will come to the consumer from the small numbers of spores in the meat. However, if the temperature of the meat is allowed to fall slowly to below 50°C, and stay warm for some time, the spores will germinate and the vegetative bacteria multiply to dangerous numbers.

Prevention

- Meat and meat products should be handled away from other foods to prevent cross contamination.
- Foods should be covered to exclude dust.
- Potatoes and root vegetables should be prepared away from other foods to prevent cross contamination from the soil adhering to them.
- Kitchen staff should maintain a good standard of personal hygiene.
- Meat dishes should, whenever possible, be prepared on the day they are to be consumed and be served hot as soon as possible after cooking.
- Where this is not possible, the meat and stock should be separated and cooled *quickly* to 10°C and refrigerated or frozen. They should be heated *thoroughly* when it is to be served.
- Hot gravy must *never* be added to cold meat.
- Joints should not exceed 3 kg in weight, to allow for efficient heat penetration.

Bacillus cereus

Bacillus cereus causes a toxic form of food poisoning.

The bacterium

Bacillus cereus is a large rod-shaped, motile bacterium which produces spores when conditions are unfavourable for its growth. It is aerobic.

The disease

Bacillus cereus food poisoning can take two forms as the organism can produce one of two toxins – but not both. If toxin A is produced in the food, the illness is similar to *Staphylococcal* poisoning. The incubation period is short – one to five hours – and the symptoms are nausea and vomiting and, occasionally, diarrhoea. The illness usually lasts six to twenty-four hours.

If toxin B is made when the food reaches the intestines, the disease resembles *Clostridium perfringens* food poisoning. The incubation period, is longer – between eight and sixteen hours, and the symptoms are diarrhoea and abdominal pain, occasionally also with nausea. The symptoms usually last 12–24 hours.

Sources of infection

The bacillus lives in the soil and is often brought into the kitchen on contaminated cereal foods such as rice and cornflour. Spices are another source of the organism.

Types of food and cooking methods involved

This type of food poisoning usually arises when foods such as rice have been partly cooked, then stored in warm conditions. This treatment creates ideal conditions for spores to germinate and the bacteria to multiply and produce toxin. The toxins are not easily destroyed by heating so warming or lightly cooking the infected food will not make it safe.

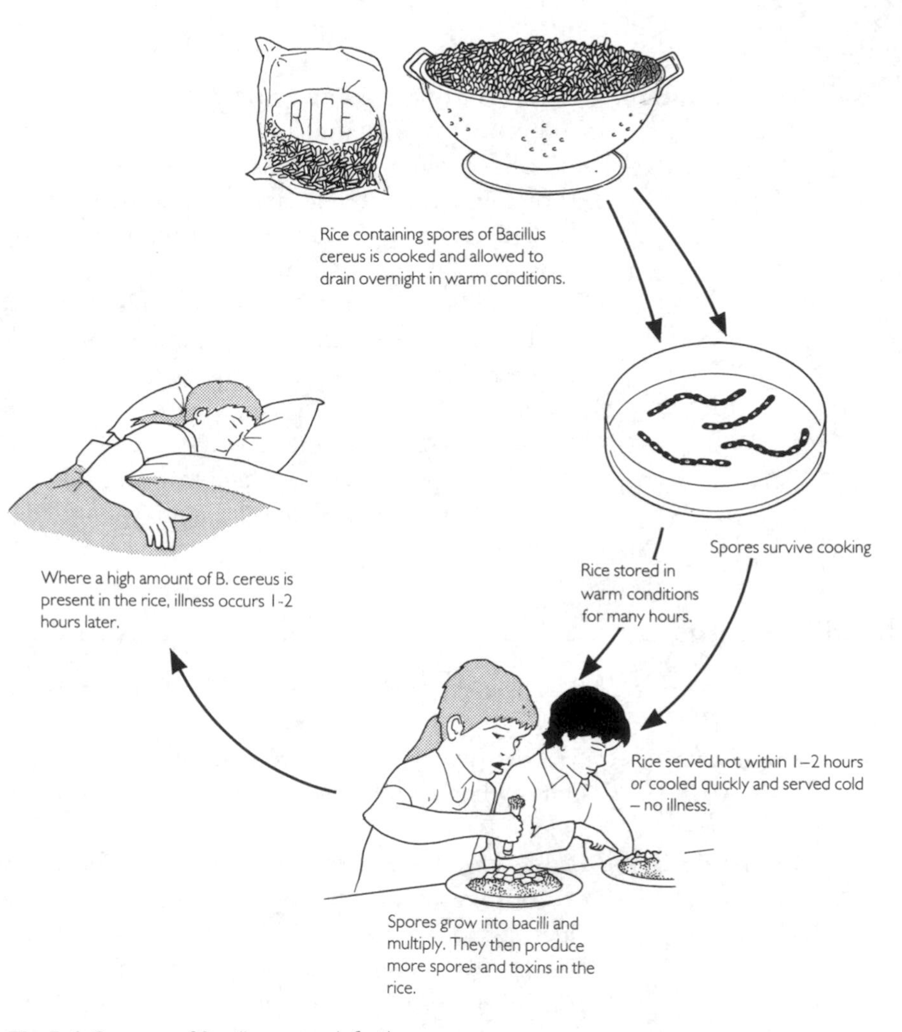

Fig 5.4 Sources of *Bacillus cereus* infection

Apart from cereal foods, the organism thrives in reheated meat dishes such as shepherd's pie if the reheating temperatures used do not prevent the spores from germinating.

Prevention

- Rice should be boiled in small batches, kept hot (above 63°C) and served as soon as possible.
- If rice is needed, cold, as for a buffet, it should be cooled quickly after cooking and placed in a refrigerator as soon as it reaches 10°C.
- If meat or rice dishes must be reheated, the heating must be rapid and thorough.
- **Never reheat** rice or meat dishes **more than once**.

Staphylococcus aureus

Staphylococcal food poisoning is due to a toxin which seeps into a food where the bacteria are actively multiplying.

The bacterium The cells are spherical and form grape-like clusters. They are not motile and do not produce spores. *Staphylococcus aureus* grows best in the presence of oxygen but can grow in anaerobic conditions.

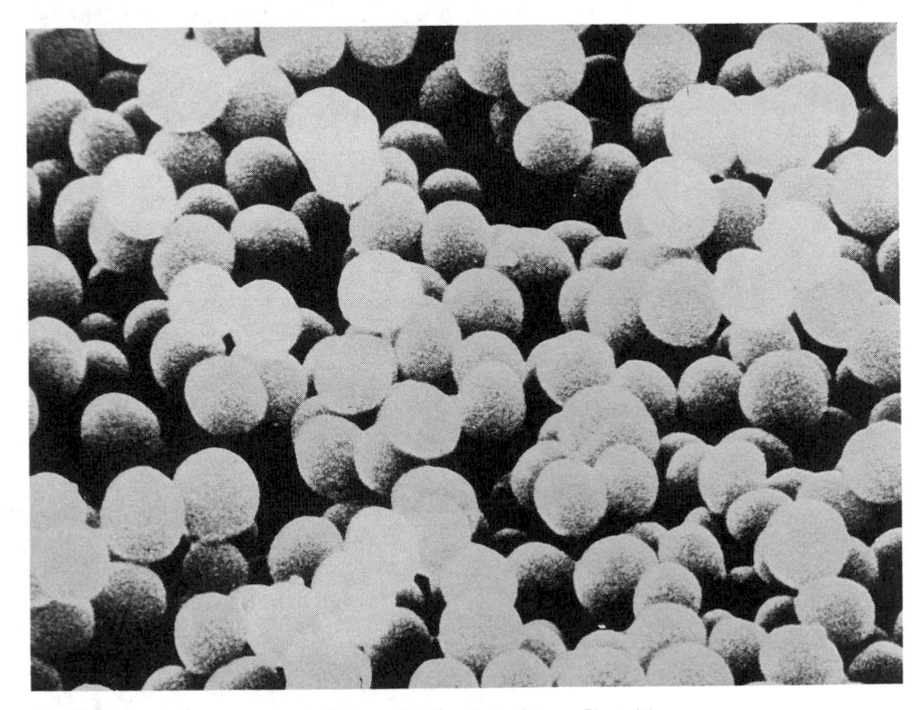

Fig 5.5 *Staphylococcus aureus Courtesy: Dr Tony Brain; Science Photo Library*

The bacteria are easy to kill by heat, they are destroyed by boiling for one to two minutes or holding at 70°C for 15 minutes, but the toxin is much more resistant. It will withstand boiling for at least 30 minutes – and longer in some conditions. This means that an infected food cannot be made safe by mild cooking methods. The bacteria may be destroyed but the toxin they have produced will still be active. *Staphylococci* are more resistant to cold temperatures than most non-sporing bacteria so may grow in refrigerators which are above the recommended 1°–4°C setting. They also survive salt concentrations which would prevent the growth of most bacteria. The ability to tolerate high salt conditions and grow in the absence of air means they can multiply in vacuum packs of bacon or preserved meats if the temperature is favourable.

The disease The toxin causes vomiting, stomach cramps and sometimes diarrhoea. As the toxin is already present in the food when it is consumed, the incubation period is short, between one and six hours. The illness usually lasts 6–24 hours. The severity of the disease can vary from one person to another, even from the same infected meal. In some people the vomiting is severe and can cause dehydration and collapse.

Sources of infection

This type of food poisoning is nearly always associated with human contamination of food. Over 50 per cent of adults carry the organism in their noses and it easily gets transferred to their mouths and hands. The organism is usually harmless when in the nose or on intact skin. However, if it penetrates the skin it can cause boils, styes, barbers' rash and septic cuts. People suffering from these conditions *should not* handle food. The disease occurs when the bacteria are allowed to multiply to dangerous numbers in a food – about one million per gram of food is the minimum level to cause illness.

Fig 5.6 Sources of *Staphylococcus* infection

Cows may suffer from mastitis, a Staphylococcal disease of the udder, so unpasteurised milk or cream may contain this bacterium.

It is a common inhabitant of dust and is easily spread from dirty equipment and grubby towels.

Types of food and cooking involved Almost *any* type of food can be the cause of this type of food poisoning, if handled unhygienically. However, the majority of cases arise from cold meats, cold sweets and cakes filled with fresh or artificial cream.

Prevention Good personal and kitchen hygiene are the best means of preventing Staphylococcal food poisoning:

- Minor cuts should be covered with waterproof dressings.
- People with septic cuts, boils, styes, conjunctivitis (pink eye) or sinus infection should not handle food.
- Keep the kitchen as dust free as possible. Use disposable towels or boil cloth towels frequently. A hot air drier is ideal for hands.
- Keep foods *covered* to prevent contamination by dust.
- Use *clean* implements to handle cold meats.
- Use disposable piping bags to decorate foods or sterilise cloth or plastic ones (*see* p 63).
- Store cold sweets covered and under refrigeration.
- Avoid coughing or sneezing near food.
- Wash your hands after using your handkerchief. (Paper tissues are preferable so long as they are disposed of hygienically.)

Clostridium botulinum

Botulism is a rare but deadly form of toxic food poisoning.

The bacterium *Clostridium botulinum* is a rod-shaped sporing bacterium. It is widely distributed in soils and is found in the intestines of animals. One particular strain (E) is found in fish in certain parts of the world (USSR, Japan and the Great Lakes area of Canada and the USA).

It is strictly anaerobic so will only grow in the absence of free oxygen, so growth and spore production are assisted by any heat treatment which drives off oxygen but is not severe enough to kill the spores. Two of the strains of *C. botulinum (A and B)*, which cause poisoning in man, are very heat resistant. They can survive boiling for hours but are killed within five minutes at 121°C.

Apart from heat, the acidity of the food also plays a major part in deciding whether *C. botulinum* can grow and produce toxin. No growth is possible below pH 4.5 (*see* p 50), so acid foods such as the majority of fruits can be cooked or canned at temperatures below 121°C without danger of causing botulism. Canned meats and vegetables which are neutral, or only slightly acid, have to be processed at 121°C, a treatment often known as a 'botulinum cook'.

The toxin

C. botulinum toxin is not heat resistant like the toxin of *Staphylococcus aureus*; it can be inactivated by heating to a full boil for 15 minutes. However, in view of the often fatal consequences of botulism, you must **never** take the risk of using a suspect food hoping that heating it will make it safe.

The disease

In warm, moist anaerobic conditions *C. botulinum* produces a deadly toxin which is absorbed by the small intestine. The toxin attaches to nerves causing gradual paralysis of the heart, digestive and respiratory muscles. It is an exceptionally poisonous substance – death has occurred as a result of eating only a few mouthfuls of an affected food.

The incubation period is long, usually 12–36 hours. The disease is difficult to diagnose because at first the symptoms mimic those of other conditions. The victim may feel giddy and suffer from headache, nausea and vomiting. By the time the victim shows unmistakable signs of botulism – double vision and difficulty in speaking and swallowing, it may be too late to save life. However antitoxins are available and, if given in time, help recovery from the disease although the process is slow, often taking 6–8 months.

Sources of infection

C. botulinum can come into the kitchen in soil on vegetables and mushrooms, in raw fish products and in defective cans of meat, fish and vegetables.

Packs of smoked fish should also be regarded as possible sources of *C. botulinum* spores as modern smoking methods tend only to impart flavour and cannot be relied on to inhibit the growth of bacteria. Smoked fish should be kept frozen in storage.

Types of food involved

In Britain, the few cases which have occurred have almost always been due to the use of commercially canned or bottled foods which escaped proper sterilisation or were later contaminated through defects in the cans.

In the USA cases have arisen as a result of people home bottling or canning meat, vegetables or mushrooms and not taking the products to a high enough temperature to kill the spores. Alternatively, in Asian countries, cases have occurred as a result of eating raw fish dishes which are often regarded as delicacies in some parts of the world.

C. botulinum spores can persist for long periods in frozen vegetables. If cooked correctly from the frozen state there is no danger from the few dormant spores in the food. If on the other hand foods are thawed out and left in warm anaerobic conditions, spores can germinate and the bacteria multiply and form toxin.

Prevention

- Discard any cans of food which show signs of the following defects *without opening them*:

Table 5.1 Characteristics of food poisoning illnesses

Cause of disease	*Incubation period*	*Symptoms*	*Duration of illness*	*Destruction of organism*	*Dangerous practices*	*Foods involved*
Salmonella infection	12–24 hours	Diarrhoea, abdominal pain, usually fever	1–7 days	No spores, easy to kill by heat. 70°C for 15 mins for most resistant strains.	Cross infection from raw to cooked foods. Poor personal hygiene. Infection from carriers. Inadequate defrosting of meat and poultry.	Meats, poultry, eggs, unpasteurised milk, meat pies and left-overs.
Clostridium perfringens infection toxin made in intestine	8–22 hours	Diarrhoea, abdominal pain, no fever	12–24 hours	Spores produced. Temperatures in excess of 100°C needed to kill spores	Meat dishes, gravies left to cool slowly. Inadequate reheating. Contamination from soil and from human carriers.	Meat, poultry, meat dishes, particularly reheated dishes. Spices often contaminated.
Bacillus cereus toxins	1–5 hours 8–16 hours	Two types:- 1 Nausea, vomiting, stomach cramps. 2 Abdominal pain, diarrhoea.	6–24 hours 12–24 hours	Spores produced. Temperatures in excess of 100°C needed to kill spores. Toxin stable to heat.	Holding cooked rice and meat dishes at warm temperatures for long periods.	Rice, cornflour, meat dishes. Spices contaminated.
Staphylococcus aures	1–6 hours	Vomiting, stomach cramps	6–24 hours	No spores produced. Bacteria killed by 70°C for 15 minutes. Toxin stands boiling for 30 mins or more	Food handling by staff suffering from skin infections, uncovered cuts. Cross infection from raw to cooked foods. Coughing or sneezing near food. Dirty towels, equipment, piping bags.	Foods handled in preparation, cold sweets, cream fillings, custards, sandwiches, unpasteurised milk or cream.
Clostridium botulinum	12–36 hours	Double vision, difficulty in swallowing and breathing	Death in 1–8 days or slow recovery over 6–8 months	Spores produced. 121°C for 5 mins needed to kill spores. Toxin inactivated by boiling for 15 minutes.	Using blown or damaged cans of food. Refreezing vegetables. Keeping smoked fish or vacuum packed fish or meat in warm temperatures.	Under heat treated or defective commercially canned meat, fish or vegetables. Home canned meat, vegetables, mushrooms. Raw fish.

1 bulging of the can or any signs of gas pressure inside
2 seams which are leaking or discoloured
3 serious denting or rusting

- Do not attempt to can or bottle any meat or vegetable – leave this to the commercial processors. Any surplus vegetables can be frozen instead.
- Do not refreeze frozen vegetables. Cook frozen vegetables directly from the frozen state.
- Keep smoked fish frozen during storage.
- Keep vacuum packed meats, fish and vegetables refrigerated or frozen before use.

Apart from *Salmonellae, C. perfringens, Staph. aureus* and *B. cereus* which are responsible for the majority of food poisoning cases, there are a number of other organisms which occasionally cause gastric illnesses. Some of these bacteria are covered in the following section.

Other food poisoning organisms

Escherichia coli

Most strains of *E. coli* live as *commensals* in the intestines of man and other animals. A commensal is a organism which lives in or on another creature and neither harm each other under normal circumstances. *E. coli* is so common in the colon or lower intestine that water analysts use it as an indicator of faecal pollution – if large numbers of *E. coli* are found in a water sample, it is taken as proof that it has been polluted with sewage.

The majority of strains of *E. coli* are quite harmless to us when they are in their normal place in the colon. If, however, they get into other parts of the body, for instance the bladder or kidney, they cause serious disease.

A few special strains of *E. coli* cause food poisoning. These are called enteropathogenic strains, meaning they cause disease in the gut. Enteropathogenic strains cause food poisoning in young children and sometimes traveller's diarrhoea in adults.

The bacterium *Escherichia coli* is a small rod shaped organism. Some strains are motile. No spores are produced.

The disease The incubation period is 12–24 hours. The symptoms are abdominal pain and watery diarrhoea. The illness lasts 1–5 days.

The organisms have been found in a wide variety of foods including mashed potato, cream cakes, creamed fish and soft cheeses.

Prevention

- Good personal hygiene particularly as regards hand washing after using the toilet helps to prevent this disease.
- Special care is needed in preparing babies' bottles and food for toddlers.

- Increasingly, cheese manufacturers are pasteurising the milk to protect their products. However soft cheeses must be handled hygienically after purchase, stored under refrigeration and used within a few days of opening.

Vibrio parahaemolyticus

Vibrio parahaemolyticus is a bacterium which inhabits fish and shellfish of subtropical and tropical countries. It is the commonest cause of food poisoning in Japan where raw fish and shellfish form a large part of the diet. It occurs occasionally in Britain and other European countries as a result of consuming imported food or infected meals on airflights from countries where the organism is endemic.

The bacterium *V. parahaemolyticus* is a curved rod with a single flagellum. It grows well in salty conditions. No spores are produced.

The disease The incubation period is 12–24 hours. The symptoms are abdominal pain and severe diarrhoea with fever and vomiting. The illness generally lasts 2–5 days.

Foods involved The organism is found *only* in fish and shellfish but grows well in other foods if they are infected by raw seafoods or contaminated surfaces. In one survey of Japanese hotels and restaurants *all* the chopping boards were found to be infected with this organism.

Prevention

- The organism is easily killed by heat. Seafoods hygienically prepared and properly cooked will be safe.
- Raw and cooked foods must be kept apart.
- *V. parahaemolyticus* survives well in ice, so airlines should avoid taking ice on board in countries where the organism is endemic.

Yersinia enterocolitica

Yersinia enterocolitica is a bacterium which came to the fore as a cause of food poisoning in the 1980s. It has been found in a wide variety of foods – milk, fresh and artificial cream, liquid egg, cheesecake, prawns and vacuum packed meats have all been shown to carry significant numbers of this bacterium on occasion.

The organism The bacterium is a short rod shaped bacterium fringed with flagella. No spores are produced. Like *Listeria* (*see* Chapter 8) the bacterium grows appreciably and is more virulent when grown at refrigerator temperatures.

The disease The patient usually suffers from diarrhoea, abscesses and abdominal pain. The symptoms can sometimes be mistaken for appendicitis.

Prevention

- Use pasteurised milk and protect it from being contaminated by dust in the air.
- As no spores are produced, thorough cooking kills the organism in foods.
- Protect foods from contamination by mice as these animals are highly susceptible to this disease.

Table 5.2 Characteristics of food poisoning illnesses

Causes of disease	*Incubation period*	*Symptoms*	*Duration of illness*	*Destruction of organism*	*Dangerous practices*	*Foods involved*
Escherichia coli	12–24 hours	Watery diarrhoea, abdominal pain	1–5 days	No spores produced. Easy to kill by heat.	Poor personal hygiene. Poor hygiene in food handling and storage.	Wide variety of foods e.g. cream cakes, soft cheeses.
Vibrio para-haemolyticus	12–24 hours	Abdominal pain, severe diarrhoea, fever, vomiting	2–5 days	No spores produced. Easy to kill by heat	Contamination of food contact surfaces by raw fish or shellfish.	Raw or partly cooked fish or shell fish.
Yersinia enterocolitica	24–36 hours	Diarrhoea abcesses, abdominal pain.	variable	No spores produced. Easy to kill by heat.	Storage at high refrigerator temperatures. Allowing food to be contaminated by mice.	Wide variety of foods e.g. milk, prawns, cheesecake.

Exercise

1 Which of the following food poisoning organisms produce spores?

(a) *Staphylococcus aureus*
(b) *Bacillus cereus*
(c) *Salmonella*
(d) Clostridium perfringens
(e) *Clostridium botulinum*

A (d), (e) B (a), (e) C (b), (d), (e) D (a), (b), (e)

2 Group the list of cooking methods under the headings A or B.

(a) Poaching
(b) stewing
(c) roasting
(d) grilling
(e) braising

A Likely to kill spores B Unlikely to kill spores

3 Which of these organisms develop a toxin in the foods they infect?

(a) *Salmonella*
(b) *B. cereus*
(c) *C. perfringens*
(d) *Staph.. aureus*
(e) *C. botulinum*

A (d) (e) B (b) (d) (e) C (a) (b) (e) D (b) (c) (e)

4 Which of the following statements is **incorrect**?

Staphylococcus aureus toxin:
(a) causes vomiting
(b) makes the patient ill very quickly
(c) is easily inactivated by heating.
(d) is often found in unhygienically handled food.

5 Match the following infections with the most likely cause:

(a) *Salmonella*
(b) *B. cereus*
(c) *C. perfringens*
(d) *C. botulinum*
(e) *S. aureus*

A Food handler with an uncovered septic cut.
B Meat casserole dish prepared too long in advance.
C Inadequate defrosting of chicken.
D Damaged can of fish.
E Cooked rice left in warm kitchen.

6 Match the food poisoning bacteria with the foods in which they are most often found:

A Poultry and eggs
B Cold meats and cream fillings
C Reheated meat dishes
D Vacuum packed and tinned fish
E Rice and cornflour

(a) *V. parahaemolyticus*
(b) *C. botulinum*
(c) *B. cereus*
(d) *Staph. aureus*
(e) *C. perfringens*
(f) *Salmonella*

7 Which of the following organisms grow well under anaerobic conditions?

(a) *C. perfringens* **(b)** *B. cereus* **(c)** *Salmonellae* **(d)** *C. botulinum*

A (a), (b), (d) B (a), (d) C (b), (c) D (b), (d)

8 Favourable conditions for growth of anaerobes would be found in:

(a) shallow fried fish
(b) vacuum packed smoked salmon
(c) the centre of a large boiled ham
(d) casseroled meat stews
(e) grilled steaks

A (b) only B (a),(e) C (b),(c),(d) D (b),(c)

9 Match the lists:

A An indicator of faecal pollution in water
B Occurs naturally only in fish and shellfish
C Grows appreciably under refrigeration
D Toxin not produced below pH 4.5

(a) *C. botulinum*
(b) *E. coli*
(c) *V. parahaemolyticus*
(d) *Y. enterocolitica*
(e) *Salmonella*

10 Match the symptoms with the disease:

A Rashes provoked in a few people after eating strawberries.
B Dizziness and breathing difficulties after eating mackerel.
C Fatality after eating a mushroom-like fungus
D Blood damage after eating wrongly prepared Mexican dish

(a) scrombotoxic poisoning
(b) red bean poisoning
(c) solanin poisoning
(d) food allergy
(e) *Aminita phalloides* poisoning

Answers

1 C (b), (d), (e)
2 A (c), (d) B (a), (b), (e)
3 B (b), (d), (e)
4 (c)
5 A (e), B (c), C (a), D (d), E (b)
6 A (f), B (d), C (e), D (b), E (c)
7 B (a); (d)
8 C (b), (c), (d)
9 A (b), B (c), C (d), D (a)
10 A (d), B (a), C (e), D (b)

6 Preparation and cooking of foods

Hors d'oeuvre, starters and salads

All dishes that are to be served cold must be handled carefully to prevent cross contamination during preparation, cooking, cooling and storage. The food must not come into contact with any raw meat or poultry after cooking or during storage.

After cooking and cooling, the food should be held in a coldroom at about 7°C, if required for service that day or in a refrigerator between 1° and 4°C if needed for the next day. None of the dishes listed above must be allowed to stand in the warm restaurant or dining room prior to service. They must be kept cool until required. Any dishes which have been on display in the restaurant/dining room must be thrown away at the end of the meal because they will have reached a dangerously high temperature. To prevent unnecessary wastage, a back-up supply of food (which has been refrigerated) should be available to replenish food on display.

Shellfish cocktails **must** be kept refrigerated until ordered. This prevents decomposition of the fish and keeps the cocktail crisp and fresh. All types of pâtés, cold fish, cold meat, poultry and game must be handled as little as possible. They must **never** be allowed to come in contact with any item of equipment which has been used to prepare raw meat or fish.

Some salad bars are constructed with tables which can hold ice to keep salads cold, others have concealed refrigerated units for this purpose. The bars are also fitted with sneeze guards to prevent people breathing on the food. All items used for raw salads must be washed thoroughly (*see* p 94) and kept refrigerated.

Stocks, soups and sauces

When making bone stock for soups, stews and sauces, **always** pay attention to the following points:

- Always use *fresh* bones and trimmings of meat and poultry.
- The stock pot must be *clean*.
- Use *fresh* vegetables for flavouring.
- Cover bones with water and bring *quickly* to the boil. Blanch if necessary.
- Simmer for the required length of time (maximum eight hours).

- Strain stock, cool *quickly* to below 10°C and refrigerate until required.
- Always date the stocks when they are made.
- Do not leave the bones in the kitchen overnight – remove them straight away to the swill area.
- *Never* leave the stock to 'cool' overnight in the kitchen.

Cooling stocks Pour the stock into shallow pans, cover and place in a cold area. When cool, refrigerate. Alternatively, place the stock in small deep pots in a deep sink of cold water until the stock reaches 10°C, when it is ready to refrigerate.

Cooling soups and sauces Soups and sauces must also be cooled as quickly as possible, dated to show when they were made, and then refrigerated. When required for use, they must be brought to the boil as *quickly* as possible. If kept hot for service, they must be held at 70°C or above. If they have been reheated but not used, they must be thrown away.

Vegetables

Vegetables are obtained from various parts of plants (e.g. roots, leaves, seeds, etc) and thus they grow:

- below the ground – root crops and tubers e.g. carrots, parsnips, beetroot and potatoes.
- on the ground – e.g. marrows, courgettes, cabbages, lettuce, celery and asparagus.
- off the ground – e.g. tomatoes, brussels sprouts, sweet peppers and varieties of beans.

All vegetables must be washed before being eaten (raw or cooked) to remove soil, sand and pesticides. If soil is left on when the vegetables are prepared, it will become embedded in the food and give it an unpleasant gritty texture.

Vegetables grown in the soil must be thoroughly washed to remove all soil, preferably before they reach the preparation area and *certainly* before they arrive in the cooking area. This will prevent any pathogenic bacteria in the soil contaminating the food.

Vegetables grown on the ground may have soil adhering to them even if they have been grown on straw. Cabbages, lettuces and celery will certainly have soil clinging to the lower leaves and this must be removed before any preparation begins.

Lettuces The varieties of lettuces fall into two main groups; those with tight heads such as Iceberg and Webbs and those with loose heads or open hearts.

The leaves of the tight headed varieties must be separated if washing is to be effective. Plenty of cold water is essential. For the loose or open hearts proceed as follows:

- Remove any outer leaves which are wilted or dirty.
- Trim the stump to a point.
- Hold the stump towards the palm of the hand and immerse up and down in the water. This will remove all dirt and insects from the leaves.
- Drain with the stalk uppermost.

After washing these lettuces are best kept whole as this helps to retain their freshness. They should be refrigerated until required.

Celery Celery which has been grown in earth must be separated into sticks and thoroughly washed to remove all traces of soil before being prepared. A small brush similar to a nail brush, which is kept specifically for the purpose, is ideal.

Other vegetables grown on the ground Vegetables grown on the ground must be washed before preparation with the exception of brussels sprouts which are prepared before being washed.

Many vegetables are washed before they are packed for both the wholesale and retail market. This ensures that soil pathogens are not brought into the preparation and cooking areas. The quality of the vegetable is easily determined because of the absence of soil and you get the weight of vegetable you order.

Storage

Vegetables should be stored in a dark, humid atmosphere at a temperature of 4°–6°C. These conditions help to retain the moisture and colour of green leaved vegetables.

After preparation, all vegetables must be refrigerated if not required for immediate use. They should be covered with damp greaseproof paper or cling film to prevent surface drying. Potatoes should be covered with plenty of water.

Fruits

All fruits must be washed before being prepared for eating raw or cooking. This applies to even to soft fruits such as strawberries and raspberries which look clean – pesticides and bacteria are invisible to the naked eye. After washing they must be well drained and kept refrigerated until required.

Citrus fruits

Citrus fruits must be washed to remove the insecticides and preservatives which have been used on their skins. This is necessary because the skins are used in making various sauces, as flavourings in cakes and fillings, and in marmalades. It is also important because any contaminants could be transferred to juice extracted from the fruit or to other foods.

Storage

Most fresh fruits should be stored at 4°–6°C but tropical fruits are usually stored at 10°C and bananas at 13°C.

Rice and pasta

Rice may be served as an alternative to potatoes, as an accompaniment to Asian or Oriental dishes, or as hot or cold salads and sweets. It is usually cooked off in large quantities in readiness for service, when it only requires reheating.

When required for accompanying savoury dishes it must be refreshed *thoroughly* and kept *refrigerated* until required for heating. This must take place as *quickly* as possible either in boiling water, if plain boiled, or fried in a little hot butter or oil. If the rice has been braised, then reheat in a little boiling stock with butter added.

Note: Under no circumstances must rice be cooked and kept warm for service and then cooled and refrigerated and reheated later – ***this is highly dangerous*** *(see p 81 – Bacillus cereus). Any hot rice left over at the end of service must be thrown away. To prevent waste, rice should be cooked in small batches.*

Rice for salads should be treated as above i.e. boiled or braised, cooled quickly and refrigerated until needed. Rice based salads should be kept refrigerated until needed and thrown away at the end of the day if not used.

Pasta shapes are cooked off in boiling water and then refreshed. They must be kept in cold water and refrigerated until needed.

Safety points

Rice and pastas must:

- **never** be allowed to cool slowly.
- **never** be kept *warm* in service – they must be either **hot** or **cold.**
- all rice kept hot for service, but unused, **must** be thrown away.

Eggs

Caterers have always had to be careful in handling eggs as there are a number of sources of infection connected with their preparation and cooking:

- Eggs are highly nutritious foods. Once the shell is broken, they are very vulnerable to infection from unclean hands or equipment and, if kept in warm conditions, provide an ideal medium for bacterial growth.
- The danger of *Salmonellae* on the outside of eggs has been recognised

for many years – hence the need for hygienic production of eggs and care in breaking the shells to prevent contamination of the contents.
- Now, as noted in the Chapter 5, a new danger has arisen from invasive types of *Salmonella* (e.g. *S. enteritidis*) which may be present *inside* eggs.

These dangers can be minimised by following a number of simple rules when cooking with eggs.

General rules

Selection and storage

- Buy clean looking eggs and store them in cool conditions, away from foods such as raw meat, which might contaminate the shells.
- Buy eggs as needed and rotate the stock.
- Do *not* use cracked eggs.

Preparation

- Wash your hands *before* breaking eggs, to prevent your hands contaminating the contents.
- Wash your hands *after* separating eggs to avoid contaminating other food or equipment.
- Make sure *all* equipment used in the preparation of eggs is removed for washing and surfaces are cleaned *before* preparing other dishes.
- Wash everything carefully, as egg sticks very firmly to whisks and other equipment.

Consumption

- Egg dishes should be eaten as soon as possible after cooking.
- If not for immediate use, egg dishes should be refrigerated and used within two days.

Raw eggs

- In the past, raw eggs have been mixed into drinks and used for such products as mayonnaise, mousses and ice-cream. Caterers who make their own products of this type should now use pasteurised whole egg, yolk or whites for these dishes.
- Remember that salad cream and mayonnaise can become contaminated *after* preparation. These products should be made in batches according to need, each day. A new batch should **never** be added to the remains of a previous batch. Any salad dressing left over at the end of the day is best discarded, but if it is essential to retain it, **refrigerate it.**

Boiled and fried eggs

To be completely safe from *S. enteritidis* infection, eggs need to be boiled until the yolk as well as the white is set (six minutes boiling). The yolk solidifies at a higher temperature than the white, (70°C) which is a sufficiently high temperature to kill any *Salmonellae* that might be present.

In the same way, when frying eggs the yolk should be basted with fat until the yolk is firm, or turned and cooked on the reverse side after the initial cooking. Serving 'sunny side up' with the yolk still liquid risks leaving *S. enteritidis* still capable of multiplying.

Warm dishes

- Warm butter sauces such as Hollandaise or Bearnaise sauce, should be made to order or in small batches so they are not left in warm conditions for long periods.
- Dishes which contain a sabayon (lightly cooked egg yolks) such as butter creams, certain sweet dishes and some sauces and soups which are finished with egg yolks and cream must be prepared carefully and preferably in small batches. Butter creams must be refrigerated until required.
- Meringues made from fresh egg whites must be dried out thoroughly if they are to be kept. Dishes containing a meringue topping should be eaten the same day or thrown away.

All dishes of the type described above involve some possible danger from *Salmonella enteritidis* infection if made with fresh eggs, even when carefully prepared. Such dishes should *not* be served to people who are particularly vulnerable to infection such as the *sick*, the *elderly* or immune suppressed *patients* in hospital.

Fish

All fish must be washed before preparation and particularly before filleting. The slime which is present on fresh fish must be removed as it contains food spoilage organisms. Grains of sand and mud may be present in the gills and these need to be washed away otherwise they may be transferred to the flesh during preparation, especially during filleting. Most types of fish are purchased already gutted which helps to prevent rapid decomposition, though this may not apply to the oily fish – herring, trout, mackerel – or salmon which may all be purchased in the ungutted state.

When you are gutting fish, clean them all first, then clear away the offal and clean down your work surface before commencing the next stage. Always have separate trays for the cleaned fish, for the offal, and for the bones which may be used later for making fish stock.

After cleaning, filleting or preparing, the fish must be placed on clean trays and refrigerated at 1°–2°C until required.

Note: Fish decomposes more rapidly than meat and other perishable foods.

Whole fish being stored prior to preparation must be kept in a refrigerator at 1°C, covered with wet ice to keep it moist and cold.

Frozen foods

A wide range of frozen foods is available. They need to be stored at or below –20°C for not longer than the time recommended by the manufacturers. The various classes of frozen foods need differing methods of defrosting before being cooked or served.

Vegetables

Vegetables are generally blanched before they are frozen and are best cooked from the frozen state in boiling, salted water. Peas and sweet corn are cooked by the time the water reboils but broad beans, brussels sprouts and broccoli may need a little longer cooking time. Vegetables which are not required immediately, should be refreshed, drained and refrigerated until required. Do not cook more than is necessary for the anticipated demand.

Meat and poultry

Frozen meat and poultry must be *thoroughly defrosted before* cooking. When using these frozen products you must think about three days ahead of your requirements. The correct method of defrosting is:

- Remove the meat or poultry from the freezer, noting the weight. *Do not remove the wrapper.*
- Place on a suitable tray, large enough to contain the drips from defrosting.
- Defrost in a refrigerator/cold room at 8°–10°C. Allow approximately 12 hours for 1 1/2 pounds of poultry e.g. 24 hours for a 3lb chicken.
- Ensure the meat or poultry is well away from any uncooked food.
- The whole or portioned pieces of meat must be *completely defrosted* before cooking.
- Cook as soon as possible after defrosting or keep refrigerated at 1°–2°C.
- The wrappers must not be allowed to come into contact with cooked food or any surfaces used for handling cooked food.

Fish

Frozen whole fish and cuts such as steaks and fillets should be defrosted in the same way as meat. Defrosted fish must be cooked *immediately* otherwise it will decompose rapidly. Individual cuts of prepared fish (coated with breadcrumbs) are best cooked from the frozen state.

Frozen shellfish must be defrosted and used immediately. Any unused defrosted shellfish must be thrown away at the end of the day.

Confectionery items

Gateaux, eclairs, trifles and mousses are high risk items and *must* be treated with *great care* during defrosting and storage.

- Defrost in a refrigerator, at 8°–10°C.
- Keep in the refrigerator and package until required.
- Keep well away from *all* raw foods, unprepared foods and only use sterilised equipment for handling.
- Only defrost what is required and throw away any unsold items at the end of the day.

Cold sweet preparations

Cold sweets may be made completely on the premises, purchased ready-made, either canned, chilled or frozen, or made up from a wide range of packet convenience foods. Many of the latter type are prepared from cold mixes which are allowed to set in a refrigerator. Some of the mixes do contain preservatives, but even so, refrigeration is essential. **Never** leave them in a warm atmosphere such as a kitchen or unrefrigerated sweet trolley in the restaurant. This type of dessert should be thrown away if not used within 24 hours of preparation.

Ice cream

Commercial ice creams must be placed in the deep freezer at –20°C *immediately* on delivery. If the product has been allowed to defrost it must be thrown away – **never** refreeze. Ice cream scoops must be kept in a sterilant which is changed *regularly* during service.

Ice cream made on the premises must be treated with great care throughout the production process. All equipment must be scalded before use. Other necessary precautions are:

- Egg yolks used for the English custard must be cooked thoroughly until it sets at 70°C.
- The cream must be very fresh and kept refrigerated.
- Any flavouring must be handled hygienically.
- The freshly made ice cream must be stored at –20°C until required.

The premises for manufacture of ice cream must be very clean and in a good state of repair. The staff must exercise a very high standard of personal and food hygiene. Some EHOs may wish to inspect the premises before giving permission for the manufacture of ice cream.

Cream products – fresh or mock

You must handle all products made with fresh or mock cream with *great care* and store them at 5°–7°C. All the equipment must be scalded and then cooled before use. Confectionery filled with fresh cream must be kept at 5°–7°C until required and must be displayed in a *refrigerated* sweet trolley.

All 'sell by' or 'consume by' dates must be *strictly* observed and any goods which are past these dates, must be thrown away. All varieties of cream must be kept in the containers in which they were delivered and refrigerated immediately on delivery at 5°–7°C.

Custard based sweets, trifles, bavarois

Trifles contain custard, cream, and sometimes jelly – all high risk foods. The custard must be cooled quickly and poured over the sponge just before setting point and refrigerated immediately. The trifles should remain refrigerated until required, any cream being piped on at the last moment. These popular cold sweets should not be placed on buffets until required.

Cream caramels, mousses, bavarois and egg custard flans must be treated as trifles and cream dishes.

Cold rice based sweets

Creoles, condés and rice moulds contain milk and cream and, therefore, require the same meticulous production as ice creams and custard based sweets. The correct storage temperature is 5°–7°C, preferably in a refrigerator specifically reserved for cold sweets.

Note: The strict regime followed in preparing cold sweets and cream filled confectionery is designed to prevent **Staphylococcal** ***food poisoning.***

Exercise

1 *Staphylococci* are most likely to by transferred to cold sweets from:
(a) raw meat **(b)** human hands **(c)** raw vegetables **(d)** cooked rice

2 Symptoms of *Staphylococcal* poisoning are caused by:
(a) ingestion of a toxin in the food
(b) multiplication of bacteria in the intestine
(c) the spores of the bacterium
(d) a toxin produced when spores germinate.

3 List 3 ways in which food handlers can minimise the chances of *Staphylococci* infecting cold sweets and cream confectionery.

4 It is always possible that one or two *Staphylococci* might get into one of these products even when carefully prepared. How could you prevent toxin production in the food even if this happened?

Answer

1 (b) human hands (or nose)

2 (a) a toxin in the food. N.B. *Staphylococci* do not produce spores.

3 Not sneezing or coughing near the food, using clean equipment and covering cuts with plasters.

4 By strict temperature control so that the bacteria cannot multiply and produce toxin i.e. store at 0°–1°C.

Food preparation equipment

There is a wide range of equipment used in food preparation which needs thorough cleaning after use if it is not to cause food poisoning by cross contamination. Machines which require special care and attention are food mixers and processors, slicing machines, bowl choppers and mincers. These pieces of equipment are used for raw and cooked meat, cream, and egg products. If small particles of raw flesh or cream are left behind as a result of inadequate cleaning, they will contaminate the next food which is processed in the machine, spreading spoilage and pathogenic bacteria from one food to another. For example, a serious outbreak of food poisoning took place in Scotland and the cause was traced to a batch of corned beef which had not been cooled properly after canning. Many of those affected had *not* eaten the corned beef but had consumed a variety of cold meats which had been cut on the *same machine*.

Piping bags

The most common use for piping bags is to pipe cream for decoration of cold sweets and gateaux. The bags may be made of nylon, cloth (calico) or plastic. The plastic type are the most hygienic because they are thrown away after use. Apart from piping cream, the bags are used for a variety of raw and cooked fish and meat dishes. You can see immediately the danger which would arise if a bag was used first for raw poultry and then fresh cream without being thoroughly cleaned and sterilised between operations.

Butchers' meat

Butchers' meat is obtained from domestic animals reared for food. The animals are slaughtered in licensed abattoirs. Strict hygiene regulations are enforced in abattoirs because the warm carcases and blood are ideal breeding grounds for bacteria. At all stages the carcase meat is checked by veterinary surgeons and meat inspectors for any signs of disease or contamination in the meat and offal. The carcases or sides are hung in a cooler at 0°C to allow rigor mortis to pass off. Meat is vulnerable to contamination by bacteria on the surface. This is difficult to prevent unless the carcases or cuts are pre-wrapped at the abattoir before being dispatched.

It is essential that meat is handled hygienically from the time of purchase to the time of cooking. All fresh meat must:

- be refrigerated immediately on delivery at 0°C.
- be handled with clean hands by personnel in clean overalls and hung with clean hooks or placed on trays or trolleys for dispatch to cold rooms.
- *never* be allowed to come in contact with the ground or with dirty equipment.

Many people insist that fresh meat must be washed before cooking.If this custom is followed, the meat must be dried *thoroughly* with clean, dry cloths or kitchen paper.

Where possible, meat should be allowed to hang in the refrigerator to allow a complete circulation of air. Trays must be placed under the cuts to collect the drips of blood and these must be changed frequently.

Meats which cannot be hung must be placed on trays and not left in the boxes or bags in which they have been delivered. The meat must not be piled on trays but kept to a depth of two inches to allow cold air to reach all the meat and prevent sweating. Space must be allowed on the trays for the blood or drips to be contained. All fresh meat must be stored at 0°C.

Preparation of meat

Meat must be prepared on suitable clean surfaces which are kept for fresh meat *only*. These surfaces must *never* be used for any cooked foods. Joints which are boned and rolled and may contain a stuffing must be prepared with extra care and cooked thoroughly. With this type of joint, the bacteria which were originally on the *outside* of the meat are transferred to the *inside* and they may not be killed if the joint is not cooked thoroughly.

All knives and butchery tools used for raw meat must be washed and sterilised after use. Ideally, they should be colour coded with the corresponding boards and work surfaces. This makes it less likely that staff will use them for other tasks.

Cooking meats

The dry methods of cookery – roasting, shallow frying, grilling and baking – ensure that vegetative bacteria such as *Salmonellae* are killed, providing the meat is well cooked and the internal temperature reaches at least 70°C. Large joints, if suitable, should be cut prior to cooking to make smaller joints, to ensure that the internal temperature reaches 70°C.

However, you should remember that the final temperature of the centre of a joint will not be high enough to kill any *spores* if they are present. It is, therefore, essential to guard against these spores germinating if the meat is to be served cold – by cooling back *quickly* and *refrigerating* until required. Joints which have been rolled or stuffed require cooking for a longer period than solid joints of the same weight to reach an internal temperature of at least 70°C.

Electric spit roasted joints and some kebabs may not reach the minimum internal temperature to kill bacteria. These methods of cooking should only be used when the food is to be consumed on the *same day* it is cooked.

The moist methods of cookery – braising, stewing and poaching – require extra care. The temperatures reached in these methods are never more than 100°C and usually somewhat below, and the cooking period may extend for two to three hours according to the quality of the meat. Slow, moist cooking drives out air so anaerobic conditions are produced in the meat.

Exercise

1 Name two species of food poisoning bacteria which thrive in anaerobic conditions.

2 Will they be killed by the cooking temperature of a stew?

3 If a stew is cooled back slowly or held below 55°C – what will happen to the spores?

Answers

1 *Clostridium perfringens* and *Clostridium botulinum*

2 No. The vegetative bacteria will be killed but **not** the spores

3 They will germinate and toxin may be produced

All meat dishes cooked by low temperature, moist methods must be kept at 75°C for service or cooled back rapidly to 10°C and refrigerated at 0°–1°C or, if facilities are available, blast frozen. Remember that most bacterial toxins are heat resistant. They are *not* inactivated if a food is heated *after* toxin has formed.

If, by reorganisation of kitchen schedules, it is possible to serve stews and other similar dishes on the day they have been cooked, this should always be done – it is *much safer* than preparing them in advance.
If this is not possible, the chilled or frozen food must be brought to 100°C as *quickly* as possible when required, *held* at boiling point for four minutes, and then *maintained* at 75°C until served. Meat steamed at 15 lbs p.s.i. must be treated as for the moist methods of cookery.

*Note: **Under no circumstances** must joints of meat and poultry (e.g. legs of pork, large turkey) be **par-cooked** one day and the cooking completed the following day.*

All cooked meats, if not required for immediate use, must be cooled to below 10°C as quickly as possible, dated and refrigerated at 1°–2°C until required.

Poultry and game

Poultry is plucked, eviscerated and trussed before being purchased by the catering industry. On delivery, poultry must be refrigerated at 0°C. When chickens are portioned for various methods of cooking, the pieces must be arranged in single layers on suitable trays, to ensure the temperature of 0°C is maintained. *All* poultry must be *thoroughly* cooked and kept at 70°C for service or cooled back rapidly to 10°C and refrigerated at 0°C.

Game Game may arrive at the establishment complete with feathers and fur, and not eviscerated. If this is the case, the plucking or skinning and eviscerating must be carried out in a separate area kept specifically for the purpose, well away from the main preparation areas. The fur of game may be infested with fleas or other insects and the feet may be dirty – birds' feathers and feet may be equally insanitary. Great care must be exercised in plucking, skinning and cleaning. The surface used for the plucking or skinning must not be used for the evisceration. The unusable offal must be carefully separated from the edible part. Game birds should be wiped out with a damp cloth and then dried and refrigerated at 0°C. Furred game should be held at this temperature also.

When handling any meat, poultry or game take care not to puncture or scratch your hands with any bones as this can cause infections such as tetanus *very quickly*.

*Note: **All** food handlers must have a course of **anti-tetanus injections** and keep them **up to date** with boosters.*

Food handlers should be aware of the danger of acquiring *Salmonella* food poisoning in the course of their work. This may arise as a result of contact with raw meat, poultry or game, by tasting food in the early stages of cooking or eating left-over food. It is more common for food handlers to transfer *Salmonellae* to *themselves*, than it is for them to infect *food* and hence their clients.

It is essential for your health that you sanitise all equipment and work surfaces after preparing flesh foods, and that you wash your hands thoroughly after work in the kitchen and before eating.

Rechauffé dishes, reheated foods

When you reheat left-over cooked food to make certain dishes, you must take extra care. High standards of hygiene and temperature control are essential to prevent growth of food poisoning organisms. The following rules must be *strictly* observed:

- All foods used must be thoroughly cooked.
- The foods must be refrigerated at 0°C–1°C and used within 24 hours.
- The food must be handled as little as possible.
- Meat and fish must be finely chopped or minced to allow reheating to 70°C+ as quickly as possible.
- All foods must be prepared on *clean* surfaces and **never** on surfaces used for raw foods.
- Raw and cooked foods e.g. raw minced beef and cooked beef must **never** be mixed and cooked together.
- Any sauces used for binding must be used cold or brought to boiling point – **never** warmed and left to cool.
- Hot and cold foods **must not** be mixed together.
- Made up dishes, fish cakes, meat croquets, ham cutlets, meat balls and Durham cutlets must not exceed one inch in thickness, otherwise they will burn before they reach the required internal temperature of 70°C.
- Foods must only be reheated **once** after cooking, and thrown away if not used. (*See* HACCP diagram on use of left over foods p 115)

Swill

Swill constitutes all the food scraps from the customers' plates as well as the peelings and inedible pieces of food which are produced during food preparation. Some items produced during food preparation can be used – these are known as by-products. Their uses are:

- Meat bones for stocks.
- Raw meat fat is clarified and used for dripping.
- Mushroom parings for flavouring sauces and soups.
- Tomato skins and seeds for flavouring fish dishes and stocks.
- Vegetable trimmings for soups.
- Parsley stalks for flavouring fish dishes and stocks.
- Fish bones for making fish stock.

Only good quality, fresh by-products must be used. They must be refrigerated until required.

All swill items must be removed from preparation areas at *regular* intervals and **never** allowed to accumulate. Swill bins must be provided and these must be kept covered at all times and removed from the kitchen areas at regular intervals. Swill must **never** be allowed to remain in or near the kitchen area overnight. The bins must be thoroughly cleaned and scalded before being returned to the kitchen.

All catering establishments must have a suitable area for storing swill bins prior to collection by contractors. This swill area must:

- be free from vermin and birds
- be kept tidy and all bins should have tight fitting lids which are in place at all times

- have a structure raised off the ground on which to stand the bins to prevent access by rats.
- have a supply of running water for hosing down, daily or when necessary.
- have a roof to keep the bins dry but allow a circulation of air.
- have easy access for collection of bins by the contractor. This should be a *daily* service.

Rubbish

Tins, jars, bottles, packing cases, plastic bags, paper and string are the main constituents of rubbish in catering establishments. These items must be placed in bins or plastic refuse bags. Sharp items such as tins, jars and bottles must be put in bins rather than plastic sacks otherwise they are a potential danger to the cleaning staff who deal with the refuse. Rubbish must be removed regularly so that it never accumulates anywhere on the premises. It should be placed in large industrial refuse bins which have tight fitting lids. Some large industrial establishments have crushing machines which compact the rubbish into neat packages.

Rubbish must be removed from premises at least twice a week and the refuse area must be scrubbed down and disinfected *regularly*. Accumulation of rubbish is not only unsightly but also a potential health hazard and fire risk.

7 Keeping foods hot

Various pieces of equipment are available for keeping food hot for service – hot cupboards, heated service counters, bain maries, carving trolleys, hot food trolleys, chafing dishes, heat lamps and plate warmers. The type of equipment used will depend on how the hot dishes were cooked.

Hot cupboards

Hot cupboards are fitted with sliding doors and may have a fitted shelf. They may be heated by gas, steam, or electricity but, whatever the method of heating employed, the cupboard must be capable of holding the food at 63°C or above (*see* Note on p 109). Hot cupboards are suitable for keeping baked dishes, roasts, and certain steamed dishes hot. They can also be used to heat plates, but they are less satisfactory than specially designed plate warmers for this purpose as the plates become excessively hot.

To clean hot cupboards after each service

1. Remove all food dishes and throw away any food left on them because it has been kept hot for the maximum recommended time. (*See* HACCP later in the Chapter.)
2. Remove any food scraps which have become lodged in the edges or corners.
3. Thoroughly wash out cupboards with hot detergent water to ensure removal of all spillage. Some of the material may be burnt on and require the use of a mild abrasive to clean away completely. (*See* Chapter 12.)
4. Rinse thoroughly with clean water and dry.
5. Thoroughly clean the sliding doors, particularly around the handles and remove particles of food from the door tracks.

Care of hot cupboards

- Make sure the doors run smoothly and close properly.
- Check the internal temperature at least once a week to ensure that it is working at the correct temperature to hold food at 63°C or above for the period required for service. Any drop in temperature must be corrected *immediately*.

- Some electric models have a regulo fitted to give a variation in temperature according to whether used as a plate warmer or to keep food hot. You must check to find out which number gives the correct temperature for keeping foods hot. The regulo should be checked periodically to see that it is working correctly.
- The exterior must be kept clean at all times. The tops of the cupboards are heated, to help in keeping dishes hot. Any spillage on these tops must be removed immediately as it soon becomes dry and difficult to remove. It also is unhygienic and looks unsightly.

Note: It is not possible to recommend a standard temperature setting for all hot food service equipment. Many factors influence the rate at which foods lose heat e.g. the nature of the food itself and the size, shape and material of the food containers. Most manufacturers will recommend a temperature setting for their particular equipment but this must always be *checked* under operating conditions to make sure all foods are maintained at a temperature of at least 63°C throughout the period of service.

Bain maries

Bain maries or water baths are heated by gas, electricity or steam and are used for keeping sauces, soups and stews hot, ready for service. The foods are kept in specially designed round or square containers which are placed in the water bath to keep hot. The temperature of the water must be sufficient to keep the food at, or above, 63°C during service.

To clean

- Turn off the heat.
- Place a suitably sized container under the tap and drain off the water. Take care because the water is very *hot*. Do not fill the container so full that it is difficult to carry by yourself or with assistance. *Always close the tap after use*.
- Refill with hot detergent water and wash the well thoroughly. It will be greasy from spillage and may contain small pieces of food which must be removed. Don't forget to clean round the edge and sides of the bain maries.
- Drain off, rinse with clean water and wipe dry.

In the bottom of bain maries there is a perforated base plate which must be removed before cleaning. The plate protects any thermostats, heating elements or steam pipes. It also keeps containers off the bottom, preventing any vibration. With this type of equipment it is essential to prevent limescale forming in hard water areas. This is one reason why they must be emptied and dried each day.

Any food left in the containers must be thrown away and the containers returned to the scullery for washing.

Hot food service counters

These may be separate items of equipment but are usually incorporated with the hot cupboards. They are heated by gas, electricity or steam. Some electric models produce dry heat rather than heating through the medium of water.

The tops of these units have specially designed food containers, made to measure. The hot foods are placed in the preheated containers to be kept hot for service. The food may be kept in some form of liquid, stock, sauce or water to prevent it drying up or burning. As with the previous type of equipment, the counter must be able to maintain the food at or above 63°C until needed for service.

To clean

- Turn off the heating and throw away all remaining food.
- Return containers to scullery for washing.
- If water bath method is used, treat as for bain maries.
- The dry units, because of their design, should not get dirty underneath. However, sometimes grease and dust particles find their way onto the heating plate which covers the elements. If cleaning is carried out daily, spillage will not become burnt onto the heating plate.
- During service any spillage on the counter must be cleaned up using a suitable cloth and a sterilant.

Heat lamps

These are situated over the hot plate or in the service area. The prepared dishes are placed under these lamps to await collection by the food service staff or the customer. The lamps must only be used to maintain the temperature of food which is already hot **never** to heat foods up. You will also see these lamps above carverys where they keep the joints hot while being carved.

To clean heat lamps

The bulbs should be switched off and wiped daily with a cloth and detergent water to remove grease and dust. After cleaning, the lamps should be dried. These measures are essential to maintain maximum heating efficiency and prevent dust and grease burning onto them.

Hot food trolleys

Heavily insulated trolleys heated electrically are used extensively in hospitals for distributing hot foods from central kitchens to the wards. They must be heated before the hot food is placed in them and kept plugged in until despatched to the wards. On arrival, they must be reconnected to maintain the temperature of the food at a minimum of 63°C.

To clean

The trolleys must be cleaned thoroughly after each service, using the method described for hot cupboards and for electric food counters. The trolleys must be checked at least once a week to ensure that the food is held at the correct temperature. Patients recovering in hospital are more vulnerable to food poisoning than the general population as their immune systems are often impaired, so extra care is *essential* to protect them from infection.

Hot carving trolleys

Many restaurants feature a hot carving trolley which is wheeled to the table for the meat to be carved to the guests' requirements. The carving base is heated by steam from a sealed water bath. The water is heated by a gas bottle or spirit lamp. The steam escaping from the water bath heats the air under the dome of the trolley. It is very difficult to maintain a constant temperature with this type of service trolley because heat is lost every time the dome is raised to carve the joint. However, as roast meats are cooked at a high temperature, kept hot only for short periods and consumed quickly the health risks are low compared with dishes such as stews.

To clean carving trolleys

- Turn off the heat, drain the water bath carefully and wash out with clean water.
- Remove any sauces and gravies and throw away.
- Wash the carving plate thoroughly with hot water and detergent, paying special attention to the meat grips.
- Rinse with clean hot water and dry thoroughly.

Chafing dishes

These items of equipment are used for hot buffets. The dishes are heated by spirit lamps, gas bottles or electricity. Some designs have a steam jacket surrounding the food container while others have the lamp in direct contact with the base of the container. The dishes must be preheated before the hot food is placed inside. They are designed for quick service and replenishing with food from the kitchen. They are not suitable for keeping food hot for one and a half to two hours unless they are designed to keep foods at 63°C.

Clean as already described for other equipment.

Table hot plates

These are made of toughened, heat-proof glass or metal and are electrically heated. They are ideal for keeping small quantities of food hot, when quick service is required. The plates must be preheated before the hot containers of food are placed on them.

Table hot plates are suitable for use when a selection of foods is being served for a short period and replenishment of hot food is available. Examples of uses are:

- Hot buffets, where deep fried scampi, meat balls, gougons of sole and other small savouries are being served.
- Buffet breakfasts, where the customers or guests help themselves from a selection of scrambled egg, bacon, sausage, tomatoes and mushrooms.

Plate warmers

These are designed to keep plates warm and must *never* be used for keeping foods hot. The maximum temperature in this equipment is far below that needed to maintain hot foods at 63°C.

Safety rule: All foods which have been kept hot by any method must be thrown away at the end of each service period.

Harsh words you may think! 'You're throwing away all the profits.' 'That would be OK for cold tomorrow.' 'You can't throw that away, we can use it for cottage pie.' These are the types of comments caterers may make when faced with this instruction – but any practising chef should have the ability to gauge, within reason, the amount of food required to be kept hot for any one service. Also, some foods are suitable to be cooked, refreshed and kept cold until required e.g. vegetables.

It is better to sell out and call dishes off the menu near the end of service time, than to *risk* reheating them again and causing food poisoning.

The Hazard Analysis Critical Control Point method of microbiological quality control

The aim of the HACCP method is to identify hazards which occur in food preparation and devise ways of controlling bacterial growth to ensure food safety.

HACCP consists of 4 stages:

1	Analysing the hazards	Contamination of foods, multiplication of pathogens and spoilage organisms, production of toxins.
2	Identifying the control points	The stages in the process where control is essential to prevent contamination by or growth of micro-organisms.

3	Devising ways to control bacterial growth	Establishing safe handling routines and safe times and temperatures for food handling operations.
4	Establishing monitoring	To verify that control measures have been carried out correctly.

Analysing the hazards

The first stage of hazard analysis concerns the foods themselves. When assessing the degree of hazard associated with different foods, remember the factors which promote bacterial growth. If a food:

- is moist
- has a neutral pH
- is not protected by chemical or thermal preservation **it is a high risk food**.

High risk foods can be broadly classified into three groups:

1 Foods likely to be contaminated with large numbers of spoilage or pathogenic organisms when they reach the kitchen. Examples include raw meat, fish and eggs.

The subsequent stages of food preparation must kill the bacteria to prevent rapid spoilage or food poisoning. Measures are needed to prevent contamination of other foods.

2 Foods which are safe, but highly vulnerable to bacterial growth if contaminated in the course of preparation. Incorrect handling and storage allows bacterial multiplication in these foods. Examples include pasteurised milk or cream, tinned meats and egg powders.

Subsequent handling must prevent contamination.

3 Foods already cooked and ready to be eaten, cold. Examples include cold meats and savouries, cold sweets and cream filled confectionery.

Handling and storage methods must prevent contamination and toxin production.

Critical control points

The next step is to consider the stages of food production where especial care is needed to prevent contamination or growth of micro-organisms. These points occur mainly when raw and cooked foods are being handled in the same kitchen or where there are changes in temperature, as when frozen foods are thawed or cooked food is cooled.

A good example is the cooking of a leg of pork weighing 4.5 kg (10 lbs):

Pork is a high risk food, commonly carrying *Salmonellae* or other food poisoning bacteria and occasionally, tapeworms or the worms which cause *Trichinosis*. To be certain of killing these organisms, the centre of the joint must be heated to 70°C. The main critical control point is, therefore, at the cooking stage.

Control methods

Having recognised the nature of the dangers and where they occur, the next stage is to decide how the hazardous operation can be *controlled*. In the case of the roast pork, the oven temperature and cooking time have to be specified so that the final internal temperature of 70°C is attained.

Monitoring control measures

Kitchens are busy places and staff changes occur, so it cannot be assumed that because safe routines have been devised, they will necessarily continue to be effective unless they are checked regularly.

In a factory, microbiological tests can be used to check that, for instance, no *Salmonellae* are being transferred from the raw to the cooked product. This type of testing is expensive and time consuming, and cannot be carried as a routine procedure in a kitchen. However, checks are possible for many kitchen hazards. In the case of the pork joint, the final internal temperature can easily be checked with a probe thermometer. The same equipment can be used to check that frozen poultry have been properly defrosted before cooking. Equally important are checks on the temperatures of cold rooms, refrigerators and freezers to make sure they are holding food at the correct temperatures.

Visual checks by senior staff are appropriate for the cleanliness of food machinery and other measures needed to prevent cross contamination.

The preparation of a dish will usually involve a series of hazards requiring control at a number of critical points. In the flow diagrams in Figs 7.1 and 7.2, the HACCP method has been applied to two common, but particularly hazardous, kitchen routines. Examine these diagrams carefully and then draw up your own HACCP flow chart for the preparation of a shellfish cocktail.

By the time you have reached the end of this chapter in the book, you should be able to attempt the case studies in Chapter 14.

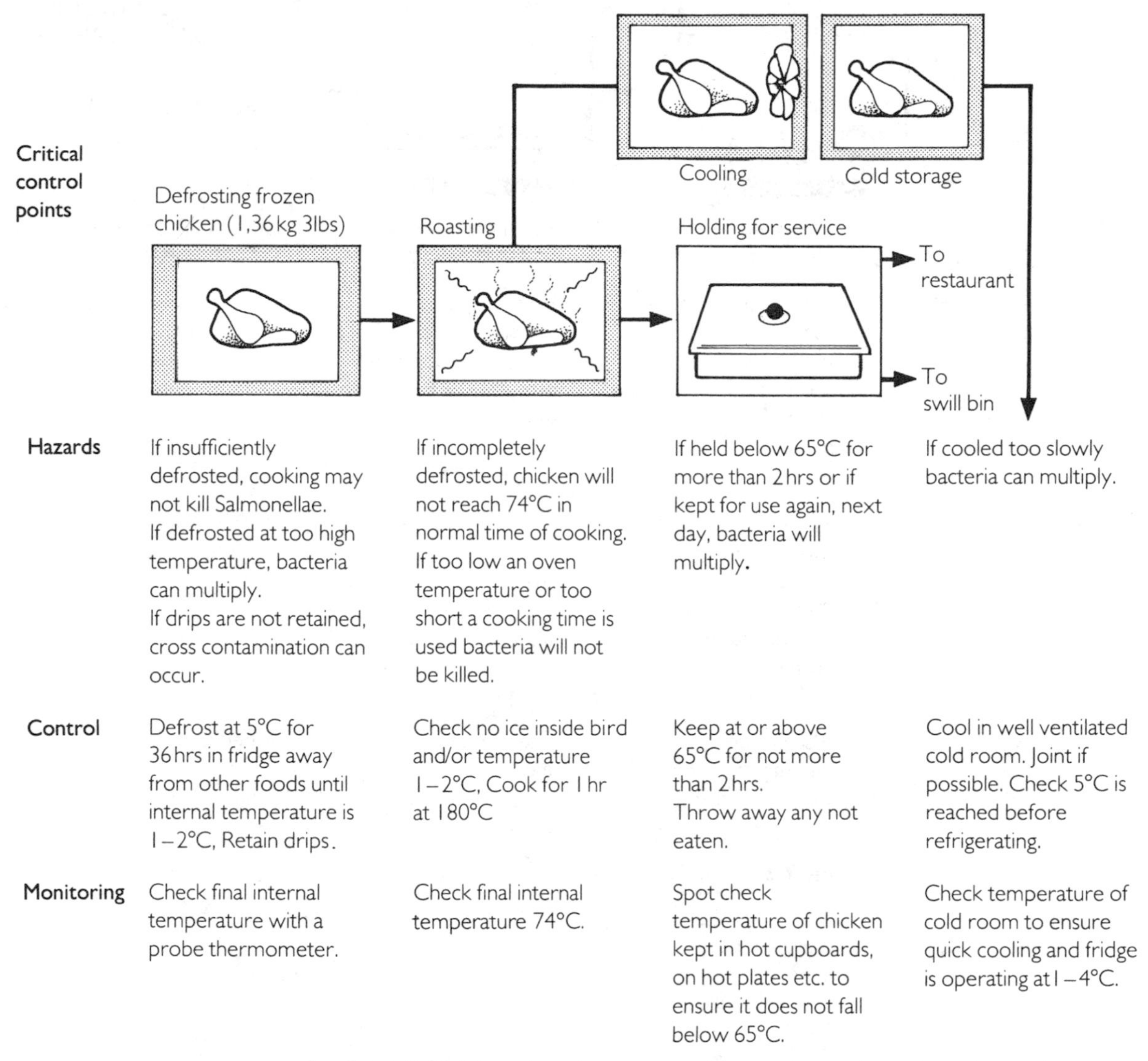

Fig 7.1 HACCP flow chart showing hazard areas of defrosting and roasting poultry

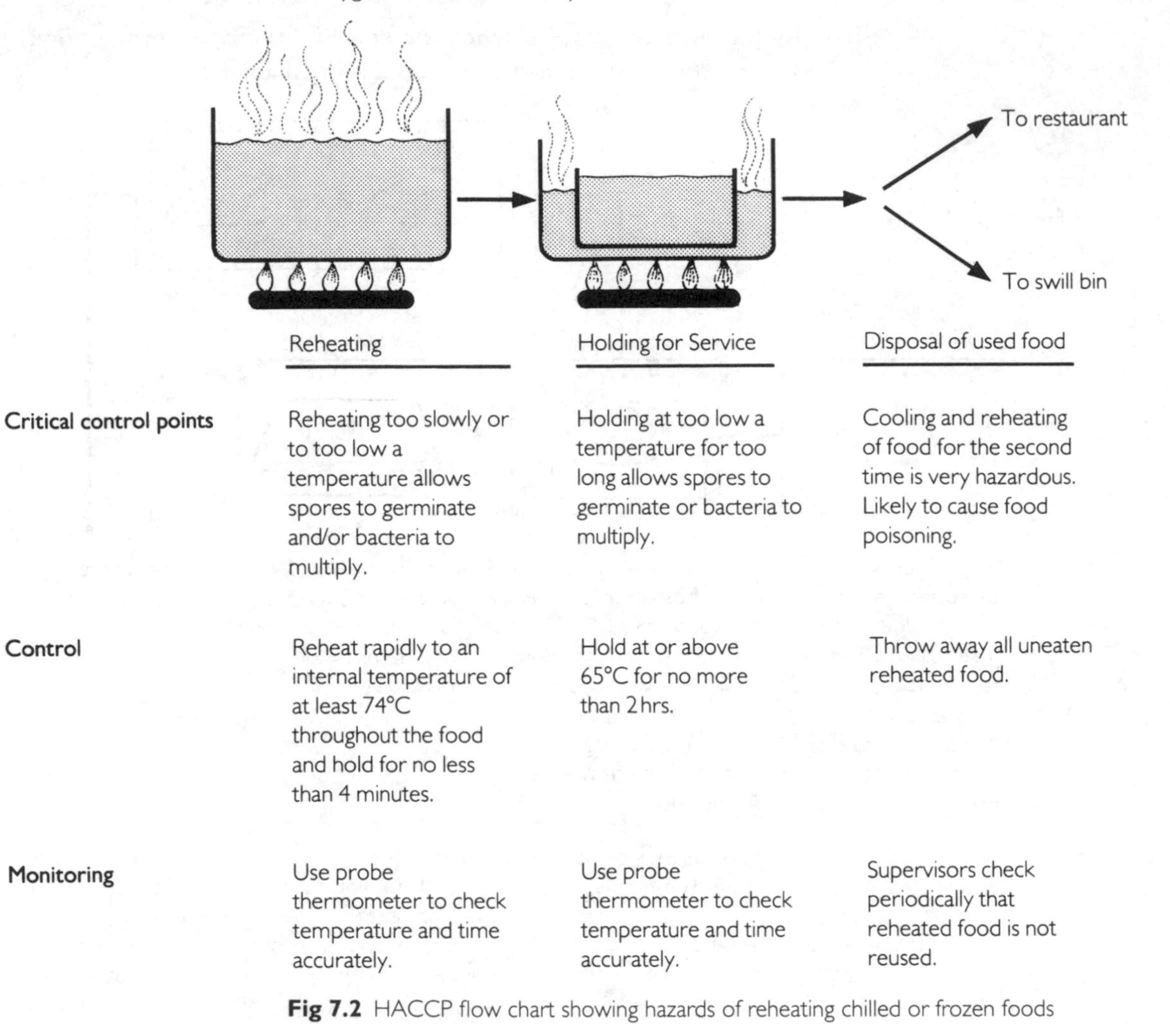

	Reheating	Holding for Service	Disposal of used food
Critical control points	Reheating too slowly or to too low a temperature allows spores to germinate and/or bacteria to multiply.	Holding at too low a temperature for too long allows spores to germinate or bacteria to multiply.	Cooling and reheating of food for the second time is very hazardous. Likely to cause food poisoning.
Control	Reheat rapidly to an internal temperature of at least 74°C throughout the food and hold for no less than 4 minutes.	Hold at or above 65°C for no more than 2 hrs.	Throw away all uneaten reheated food.
Monitoring	Use probe thermometer to check temperature and time accurately.	Use probe thermometer to check temperature and time accurately.	Supervisors check periodically that reheated food is not reused.

Fig 7.2 HACCP flow chart showing hazards of reheating chilled or frozen foods

8 Food-borne disease

Chapter 5 covered the bacteria associated with the common forms of food poisoning in the UK. However, there are many other organisms which infect drinking water and food from time-to-time and cause illness in man. These illnesses are termed food-borne diseases and differ from food poisoning in a number of ways:

- **The size of the infective dose.** – Only a few of these organisms are necessary to make people ill. In many cases the organism does not multiply outside the body – the food or drink just acts as a carrier.
- **Long incubation period.** – Many days or weeks may elapse after the initial infection before the symptoms of the disease appear.
- **Water often the source.** – Infected water supplies are often the source of these diseases. Water is never the cause of food poisoning.
- **Parts of the body affected by the disease.** – Food-borne diseases are not confined to the gut; other organs are frequently affected.

However, the division between food poisoning and food-borne diseases is not absolutely clear cut. Some people, for instance regard, *Campylobacter* to be a food poisoning organism rather than a food-borne one.

Food-borne diseases are common in countries where poverty, war or natural disasters prevent effective water treatment and proper sewage disposal. However, these diseases can be difficult to prevent, even in developed countries, since temperature control alone is not effective – the organisms do not have to multiply in the food or drink to cause illness. Only careful hygiene and inspection throughout the food chain, from farm to consumer, is likely to be successful in reducing the toll from food-borne diseases.

Campylobacter jejuni

Campylobacter jejuni has only been recognised as a major cause of human food-borne disease since 1977, but is now believed to be the most common cause of intestinal infection in the UK today. The disease, though unpleasant, is comparatively mild and has caused very few deaths.

The bacterium

Campylobacter jejuni is a thin, spiral or curved bacterium which grows best under reduced oxygen conditions. It is motile, having a tuft of flagella at each end. No spores are produced.

Table 8.1 Consumer protection in the UK food chain

Stage of food chain	*Legislation*	*Action*
Agricultural supplies		
Animal feed Fertilizer Pesticides Veterinary medicines	Food Safety Act 1990 Agriculture Act 1970	Controls the content of animal foodstuffs. All additives must be harmless to the animals concerned and in human food. Production of protein feeding materials controlled to reduced *Salmonella* contamination.
	The Medicines Act 1968	Controls the manufacture, sale and supply of veterinary medicines to prevent risk to the animals and to the consumer.
	The Food & Environment Protection Act 1985	Controls the sale, supply, storage and use of pesticides.
	The Control of Pesticides Regulations 1986.	States the maximum level of pesticide residues permitted in foods.
Primary producers Sheep Cattle Poultry Fish	Food Safety Act 1990 The Animal Health Act, 1981	Applies measures to prevent animal diseases being passed on to man e.g Tuberculosis, Salmonellosis, Brucellosis: • Regular testing cattle for T.B. and *Brucella* infection. • Testing laying and breeding hens for *Salmonella*. Slaughter of flocks with 'invasive' *Salmonella*. • Control of fish landings and sales after chemical spills.
Primary food processing		
Milk	Food Safety Act 1990 Milk and Dairies Regulation, 1959	Sets down standards for hygiene in milk processing. Lays down times and temperatures for pasteurising all milk sold to the public apart from 'Green top' milk.
	The Milk (Special Designation) Regulations 1977.	Green top milk is subjected to more stringent standards. Must be clearly labelled as raw milk.
Animal carcasses	The Slaughterhouses Act 1974. The Slaughter of Animals (Scotland) Act.	Slaughter houses have to be licensed by the Local Authority and are subject to inspection by Environmental Health Officers. Carcasses must be individually inspected to prevent unfit meat reaching the public. No offals can be used from B.S.E. affected cattle.
Food manufacturing and processing		
Canning Freezing Vacuum packing	Food Safety Act 1990	Controls all production of food to ensure that it shall be fit for consumption and correctly labelled. Manufacturers enforce their own hygiene controls through HACCP systems to comply with the Act.
Food colour tests	The Food Acts 1984.	Lists the additives permitted in foods and the maximum levels allowed. Recommends methods and sets hygiene standards and temperatures for cook-chill and cook-freeze operations.
Packet of food showing label with nutritional information	Food Labelling Regulations 1984.	Ensures that labels provide sufficient, understandable and accurate information on the contents of food for consumers to know what they are buying.
A food being wrapped or fed into a container	The Materials and Articles in contact with Food Regulations 1987.	Controls the nature of packing materials and containers so they do not transfer harmful substances to foods.
A food being irradiated		Sets up supervision and inspection of plants for irradiation of foods. Ensures that irradiated foods are clearly labelled to allow the consumer to use or reject this type of food.

Table 8.1 Consumer protection in the UK food chain (*continued*)

Stage of food chain	*Legislation*	*Action*
Food retailing and distribution		
Refrigerated van delivering at a supermarket Supermarket – shelves and freezer cabinets Market stalls selling food.	Food Safety Act 1990. The Food Acts 1984. Food Hygiene (Amendment) Regulations 1990. Food Hygiene (Market Stalls and Delivery Vehicles) Regulation 1966.	The Amended Regulations control the temperature settings of freezer and chill cabinets in food outlets. Enforces high standards of hygiene. Food should not be unfit for human consumption or falsely labelled. They must be of the nature, substance and quality demanded by the purchaser. The new Food Safety Act makes it an offence to be in *possession* of unfit food not just to sell it.
The catering industry		
Food being prepared in a commercial kitchen Food being served in a snack bar Food being collected from a take-away.	Food Safety Act 1990. The Food Acts 1984. The Food Hygiene (Amendment) Regulations, 1990.	The new Food Safety Act enforces registration of food premises. Notices can be served on any premises which are below standard. The new Act makes provision for training of food handlers in food hygiene.

The disease The infective dose is small so the organism does not need to multiply to any extent to cause illness. The incubation period on the other hand is long – 2–7 days. Patients have influenza-like symptoms at first with fever and abdominal pain. Diarrhoea follows later, which may be watery and blood stained and, sometimes, severe. The symptoms usually last 5–7 days, sometimes longer.

Sources of infection *Campylobacter jejuni* is found in cattle, pigs, poultry and domestic animals. It is rarely spread from one adult to another but it may be passed from mother to child and often spreads from one infant to another.

Types of food involved Most cases of *Campylobacter* infection have been traced to unpasteurised milk or to undercooked chicken.

Prevention Good standards of food and personal hygiene are essential to control this infection.

Campylobacter is easily killed by heat. If frozen chicken is completely thawed and then thoroughly cooked there will be no danger of infection. Pasteurisation eliminates the infection from milk supplies.

Listeria monocytogenes

Listeriosis has long been known as a disease of sheep, cattle and goats. The affected animals excrete the bacteria in their milk and their faeces. The organism survives in silage fed to animals and in the sewage sludge that is sometimes spread on farmland as a fertiliser. *Listeria monocytogenes* is widespread in the environment, in the soil and water, and is therefore capable of contaminating plants grown for food.

Pinpointing the foods responsible for listeriosis in human beings has been difficult since the incubation period is lengthy and the number of people affected in each outbreak is usually small. However, the number of cases has shown a marked increase from around 40 annually in the 1970s to about 300 in 1988. Anxieties about this increase have prompted investigations which have identified the main food groups involved with listeriosis to be dairy products, prepacked salads, precooked chicken, chilled ready-cooked chicken dinners and meat pâtés.

The bacterium

L. monocytogenes is a small, rod shaped bacterium. It has flagella, so is motile. It does not produce spores.

Listeria has a number of peculiar characteristics which make it difficult to control in foods:

1 *A wide growth temperature range.* It is capable of growth over a wide range of temperatures from 0°C to 45°C, though between 0°C and 3°C growth is slow.

This means that the organism will grow at the chill temperatures often found in many domestic and commercial refrigerators. In addition, it is actually *more* pathogenic when growing under these cool conditions than at room temperature. This means that the presence of the organism in foods which are likely to be kept chilled for several days is *extremely* hazardous.

2 *Effect of butter fat. Listeria* has another peculiarity which gives it an advantage in dairy foods. The presence of butter fat enhances growth at low temperatures and also increases pathogencity, the higher the butter fat in the product the greater the effect on the organism. This means that *Listeria* is more dangerous in full cream milk and full fat cheeses than in low fat dairy products stored at refrigerator temperatures.

3 *High thermal death point.* Finally, *L. monocytogenes* is more difficult to kill by heating than most non-sporing bacteria. Most authorities believe that the organism is killed by pasteurising provided that the temperatures and times are strictly maintained (63°C for 30 minutes or 71.7°C for at least 15 seconds). However, many American dairies have recently raised their pasteurising temperatures to 76.6°C in response to worries over *Listeria* voiced by their customers.

The disease

The incubation period is long, anything from 5 days to 6 weeks. The symptoms of the disease vary according to the age and health of the people affected. Only about ten per cent of cases of Listeriosis occur in previously healthy adults. Those most at risk are the very young, the elderly and those with weakened immune systems.

The bacterium produces a substance known as *listeriolysin*, when grown at low temperatures, which damages macrophages, the white cells of the blood which defend the body against infection. When the organism gets into the bloodstream it causes a feverish influenza-like illness. In those with lowered immunity it produces more serious conditions. The bacteria may multiply rapidly in the blood causing blood poisoning (septicaemia) or invade the membranes covering the brain (meningitis).

Listeriosis is particularly hazardous if contracted in pregnancy since it frequently causes miscarriages and stillbirths. Babies may also be born severely affected by the disease.

Listeriosis has an overall mortality rate of around 30 per cent so, although it is comparatively rare disease compared with *Campylobacter* or *Salmonella* infection, it must be of concern to the health authorities and to those who handle food.

Sources of infection

Listeria is widespread in the environment. It is found in the gut of cattle, sheep and goats and is also carried by about 15 per cent of healthy people.

Types of food involved

Outbreaks of listeriosis have been traced to milk and soft cheeses, particularly those with a high fat content such as Brie and Camembert. Salads based on lettuce and cabbage have also been implicated, particularly those prepacked and held under refrigeration. The nut and bean shoot types of salad have not proved to be infected. The organism has also been found in fresh and precooked poultry, precooked chilled poultry dishes and in meat pâtés.

Prevention

- Use pasteurised milk and cream. When bottles have been opened, protect the contents from dust.
- Keep cheeses covered to avoid cross contamination
- Pregnant women, the elderly, and those with lowered immunity should avoid eating soft cheeses and meat pâtés.
- Cook fresh chickens so that all parts reach at least 70°C, preferably 75°C.

Precooked meals

Since the largest number of *listeria* cases in the late 1980s arose from chilled ready-made meals, special care must be taken when dealing with this type of food. Precooked meals are widely used in public houses and fast food outlets, where microwave ovens are often used for their regeneration. These ovens are intended *to heat* the meals and *not to sterilise* them.

It is important to realise that several types of ready-made meals exist which have received different types of treatment in the factory:

- The safest and the easiest to store and prepare are those which are already sterilised and only require warming before service. The sealed packs can be stored on the shelf in the same way as a can of food and will keep for a year. Examples of these products include Microchef, Mark & Spencer Long Life, Boots Shapers and John West Micro-ready meals.

- Less safe and more difficult to deal with are chicken dinners and other ready-meals which have just been pasteurised in the factory and kept chilled on the way to the customer. If these products are used, it is essential to take great care to prevent multiplication of any *Listeria* bacteria which might have escaped the pasteurisation process. The following precautions are recommended:

- Arrange to have these products supplied at frequent intervals – daily if possible.
- Check 'use by' dates before accepting the products.
- Check the thermostats of refrigerators used to store these meals. Do not allow the storage temperature to rise above 3°C.
- Do not store for more than 2 days before use.
- Follow microwaving instructions on the packet *precisely,* including the 'standing time'. (*See* p 67.)
- Use a microwave thermometer to check that the food reaches a minimum of 70°C during cooking.

Typhoid and Paratyphoid

Typhoid and Paratyphoid are diseases spread mainly through faecal contamination of water supplies. Today, in Britain, the majority of cases are found in immigrants from third world countries and people who have contracted the disease when on holiday or business trips abroad. However, outbreaks of the enteric fevers (typhoid or paratyphoid) were common in Britain in the nineteenth century. Leaking sewers and poor drainage systems meant that drinking water was often contaminated with the bacteria which cause these diseases. Even royalty were not immune. The Prince Consort died of typhoid fever and Queen Victoria was said to blame the poor state of the drains at Windsor Castle for her husband's death.

Salmonella typhi

Salmonella typhi is a small rod shaped bacterium. It has a complete fringing of flagella, so is motile. It does not produce spores.

The disease The incubation period is usually 7–12 days. The early symptoms of the disease are headache and weakness. The patients later become feverish with temperatures climbing in step-wise fashion until they reach a peak of 103°F or 104°F (39.4/40°C). Rose coloured spots appear on the chest and abdomen as the bacteria multiply in the bloodstream. Diarrhoea is usual at this stage in the illness, sometimes slight, in other cases very severe. The fever continues until the fourth week of the illness and then there is a gradual return to a normal temperatures in patients recovering from the disease.

Typhoid fever has a death rate of 2–10 per cent though antibiotics have made the many complications of the disease easier to control than used to be the case.

Carriers About 30 per cent of typhoid cases become temporary carriers, excreting the bacteria in their faeces or urine for some weeks or months. About 5 per cent remain *long term carriers*. They feel healthy, but retain the bacteria in their gall bladders or other parts of their bodies and excrete small numbers of *typhoid* bacilli permanently. *Note*: Known carriers of typhoid or paratyphoid are *not permitted to work* in the food industry.

Paratyphoid fevers

The disease caused by *Salmonella paratyphi* is similar to typhoid but the symptoms are generally milder. The bacteria have similar characteristics and are spread in the same, way, but the incubation period is shorter, 1–10 days.

There are three strains of *S. paratyphi* which can be distinguished by laboratory tests. The paratyphoid fever in Britain is almost always due to the B strain, while A and C strains are likely to be picked up by people travelling abroad. People planning to travel to some Mediterranean countries are advised to be immunised with TAB vaccine to protect them from Typhoid and Paratyphoid A and B.

Sources of infection of typhoid and paratyphoid

- Water contaminated by sewage.
- Food or milk supplies contaminated by a carrier.
- Shellfish or watercress raised in contaminated water.
- Flies carrying the bacterium from infected faeces to food.
- The use of untreated sewage as a fertilizer on land growing vegetables.

Prevention

- Purification and chlorination of drinking water.
- Inspection and maintenance of sewers and septic tanks to prevent contamination of water for humans *and* from farm stock.
- Raising shellfish and watercress in non-polluted water.
- Cleaning shellfish by flushing them with clean water for 48 hours before use.
- Checking food handlers to exclude carriers.

Dysentery

There are two types of dysentery:

- Bacillary dysentry, which caused by a bacterium
- Amoebic dysentery, due to a protozoan (a single-celled animal).

Bacillary dysentery

Bacillary dysentry is a world-wide intestinal disease. Outbreaks are regularly reported in Britain and have been increasing in number since 1950s. Typical outbreaks occur in closed institutions such as homes for children, the elderly, and the mentally handicapped.

The organisms which cause the disease belong to the *Shigella* genus and resemble the *Salmonella* group in many ways.

The bacterium *Shigellae* are short rods. They are non-motile and non-sporing.

The disease The incubation period is 1–7 days. The symptoms are diarrhoea and fever, sometimes accompanied by vomiting. The severity of the illness varies greatly according to the species of *Shigella* responsible – from very severe, to mild conditions which hardly inconvenience the patient.

In tropical countries, contaminated water is the main reason for the spread of the disease but in Britain it is mainly spread from hand to mouth. Large outbreaks tend to occur amongst the inmates of institutions such as prisons or homes for the mentally retarded where satisfactory personal hygiene standards may be difficult to maintain. Inadequate toilet facilities and crowded conditions increase the incidence of the disease. Like other faecal bacteria, *Shigella* can be passed from one person to another as they flush the lavatory, touch a toilet door handle, or use the wash basin taps.

As with *Salmonella* infection, sufferers of bacillary dysentry often become temporary carriers of the disease, and their presence in a community makes control of the disease difficult.

Prevention

- Good education in personal hygiene, particularly in hand washing after using the toilet, is the most important measure in preventing and limiting outbreaks.
- Adequate toilet facilities are essential with wash basins *inside* the toilet cubicle to avoid contamination of door handle.
- Toilet facilities can be designed to reduce cross infection. The flush of the lavatory, wash basin taps, and toilet room doors can all be foot operated.
- Thorough treatment of water supplies helps to limit the spread of the disease.
- Regular inspection and maintenance of sewers and drains prevents contamination of water supplies.

Brucellosis

Brucellosis is a disease of cattle and goats which can be passed on to people through the drinking of infected raw milk, or direct contact with infected animals.

The organisms Bacteria of the *Brucella* genus are oval to rod shaped in appearance. They are non-motile and non-sporing.

The disease Brucellosis is also known as Malta fever and undulant fever.
The incubation period is long, 1–4 weeks. The symptoms vary but are usually pain in the muscles and joints with a fever which rises and falls (undulant). The disease is rarely fatal but can cause prolonged ill health.

Prevention

- Milk is made safe by pasteurisation.
- Britain has undertaken a programme of eradication of *Brucella* from cattle so that increasingly large areas of the country are now stocked with Brucellosis-free herds.

Tuberculosis

Tuberculosis is caused by bacteria of the *Mycobacterium* genus. There are two main forms of the disease:

- Pulmonary tuberculosis, which affects the lungs and is spread from person to person by droplet infection.
- Glandular tuberculosis which affects the lymph glands, joints and bones and is generally contracted through drinking milk infected with *Mycobacterium bovis* from cows.

Glandular TB was a common disease in Britain before World War II, so common in fact, that the American forces took the precaution of bringing their own milk supplies when they were stationed here.

The bacteria *Mycobacteria* are rod shaped, non-motile, non-sporing organisms. They are more resistant to heat and chemical disinfectants than most non-sporing bacteria. For this reason the times and temperatures used for pasteurisation of milk have been chosen specifically to kill *Tubercle* bacilli.

Glandular tuberculosis

The incubation period is long, 4–6 weeks as a rule. The bacteria settle in the lymph glands of the neck region or in the lymph nodes of the intestine. The disease often spreads to other parts of the body – to the joints, bones and kidney.

Prevention

- Pasteurisation of milk supplies was the initial step taken which reduced the toll of this disease.
- A programme of irradification of TB in cattle was started in Britain shortly after World War II. Cows were tested with Tuberculin prepared from the bacilli and any reacting to the substance were slaughtered. The testing continued until the whole herd could be certified free of the disease or Tuberculin Tested (TT). The measures have been highly

successful, reducing the toll of glandular tuberculosis from 1500 deaths in England and Wales in 1940 to only 5 reported cases in the years 1981 to 1983.

Viral diseases – Hepatitis A

The word hepatitis means disease of the liver. Viruses cause a number of different kinds of liver disease but only one, Hepatitis A is spread by food and drink.

The disease has a lengthy incubation period of about 4 weeks. The virus becomes established in the intestine so the faeces of the sufferer become heavily infected with the virus. Patients suffer from nausea, fever and abdominal pain and later, jaundice. Their skin becomes yellow as bile pigments from the diseased liver colour the blood.

The severity of the illness varies greatly, some people being mildly affected and recovering in a week and others severely ill and remaining in poor health for many months. Children are often less affected, generally having intestinal symptoms but no jaundice. However, they gain a lifetime immunity as a result of the mild illness.

The carrier state is common. At any one time there are many more *carriers* of Hepatitis A in the population than *cases* which show symptoms.

Sources of infection

The virus may be acquired from:

- sewage-polluted drinking water
- milk or food infected by a carrier with poor hygienic standards
- shellfish raised in polluted waters.

Prevention

- Effective sewage treatment and purification of water supplies are essential to limit the spread of this disease.
- Shellfish must be raised in clean waters.
- Good standards of personal hygiene are needed amongst food handlers, particularly as regards hand washing, to ensure that the virus is not passed to food and drink.

Hepatitis B

Hepatitis B is a more serious viral disease, but it is spread via blood, so is of more concern to health workers than to caterers.

Protozoan diseases – Amoebic dysentery

Amoebic dysentery is a common disease in many tropical and sub-tropical countries. It is caused by a protozoan, *Entamoeba histolytica*. This single-celled animal is similar to the amoeba studied in school biology but, unlike the harmless freshwater protozoan, it is a parasite in the human intestines.

The amoebae are spread from person to person in the form of cysts – round resistant bodies shed in the faeces of those suffering from the disease. The cysts are difficult to kill in water supplies so the amoebae easily pass from one person to another by this means.

The incubation period is variable and can be as much as 60 days. Inside the host, the amoebae break out of their protective cases and burrow into the intestine wall causing ulcers to form. The diarrhoea caused by the organism may be comparatively mild or may be severe and blood stained.

Prevention

- Prevention lies with obtaining protected drinking water supplies from deep wells or other sources which are not open to faecal pollution.
- Food must be protected from flies which can carry the organism.
- Good personal hygiene is necessary to prevent food handlers passing on the organism to foods.

Protozoan diseases – Cryptosporidium parvum

Cryptosporidium is a parasite whose importance has gradually emerged during the 1980s. It has world-wide distribution in animal wastes, so is a potential danger to water and food supplies.

The organism

Cryptosporidium belongs to the protozoa and lives in the intestines of farm animals such as cattle and sheep but also in domestic pets and many wild animals. The infection is passed to man by oocysts, round resistant bodies which contain motile cells of the parasite. The oocysts pose a particularly difficult problem in water supplies as they are not damaged by the chlorine used to kill bacteria.

The disease

The parasite causes an acute diarrhoea which may continue for long periods. The incubation period is variable – three days to a week as a rule, but it can be longer.

The infective dose is small. It is thought that as few as 10 to 100 oocysts are sufficient to start an infection.

Sources of infection

Cryptosporidium spreads in a variety of ways:

- from person to person as in other diseases
- by direct contact with infected farm or domestic animals, or with slurry or sludge spread on agricultural land
- by drinking or swimming in infected water
- by consuming foods contaminated by infected water or by carriers of the disease.

Amongst the foods which have been reported as causing *Cryptosporidium* infection are sausages, pâtés, offals and raw milk.

Prevention

- *Cryptosporidium* is difficult to detect in water supplies. Research is needed to improve methods of detection and destruction so that the organism can be eliminated from the food chain.
- Good personal hygiene minimises the chance of carriers passing the parasites to other people, to food or to water supplies.
- Normal cooking temperatures for foods such as sausages and meats are sufficient to kill *Cryptosporidium*. Cases of the disease have occurred as a result of picking at foods during cooking or through undercooking meat dishes. These practices should be avoided.

Worms

Foods can transmit larger pathogens as well as micro-organisms such as bacteria and viruses. A generation ago, tapeworm infestation was common amongst children in Britain but now, with better sanitation and meat inspection, these cases have become rare. Threadworms, however, are still a problem with children.

Tapeworm infestations

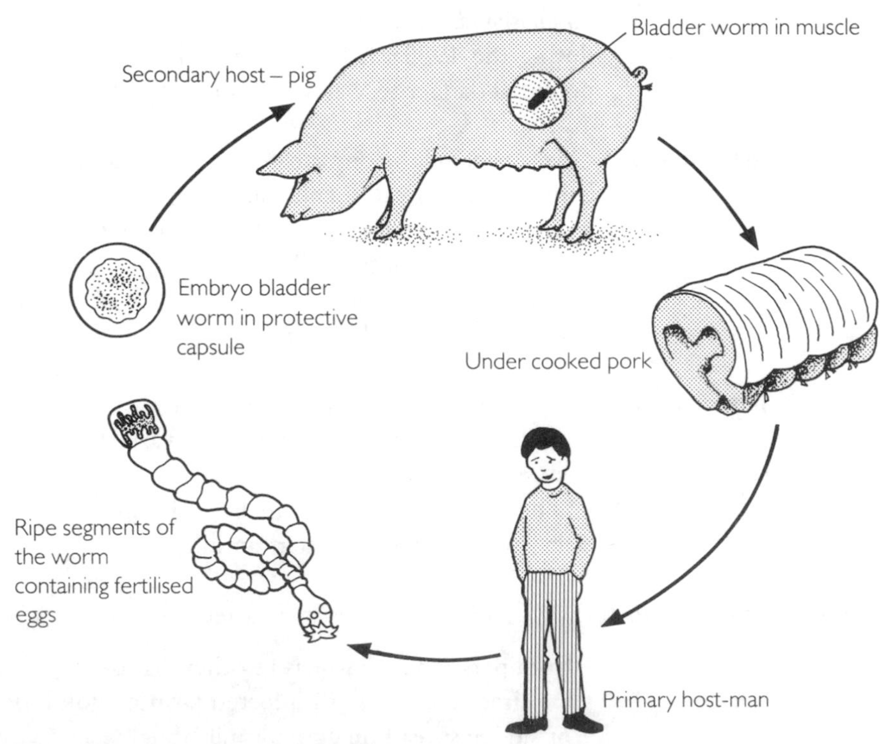

Fig 8.1 Life cycle of the pork tapeworm *Taenia solium*

These diseases are caused by eating 'measly' pork or beef containing the pork tapeworm, *Taenia solium* or the beef tapeworm, *Taenia saginata*. The bladder worms which give the meat the measly appearance contain the early stage of the tapeworm. In the human intestine, the bladder worm turns inside out to expose the small head with its row of hooks and four suckers. The head attaches firmly to the lining of the intestine and the body grows, absorbing the predigested food through its skin. The worm adds many flat, identical segments until it attains a considerable length, sometimes several feet. The ripe sections at the far end of the worm are shed with the faeces. They contain the next generation of the worm which has to reach a cow or a pig to start the cycle once more.

Symptoms The symptoms of the disease are abdominal pain and increased appetite. There is also loss of weight because the worm absorbs much of the food which would otherwise nourish the infected person.

Prevention Tapeworm infestation is a rare condition in Britain today because:

- Meat is carefully inspected to prevent measly pork or beef reaching the public.
- Sanitary conditions have improved even in remote villages. 'Night soil' i.e. human faeces are no longer spread on farmland so the bladderworms cannot reach their primary hosts to complete their life cycles.
- The public are aware of the importance of cooking meat, particularly pork, thoroughly.

Trichinosis

Trichinosis is caused by a small roundworm called *Trichinella spiralis*. It is common in the USA but occurs only occasionally in Britain.

The disease is contracted as a result of eating inadequately cooked pork containing the cysts of the worm. The protective cases of the cysts are digested along with the food and the small worms are released to burrow into the wall of the intestine. Here they grow, mature and produce many more of their kind. The worms then travel through the blood vessels to the muscles where they form cysts and begin the cycle over again.

The disease The incubation period is variable – from as little as one day to several weeks. Initially the symptoms are similar to those of food poisoning: diarrhoea, vomiting, abdominal pain and fever, but later there is muscular pain and soreness when the worms encyst in these tissues.

Prevention

- Good standards of meat inspection help to prevent much of the affected meat reaching the market, but it is not a complete safeguard as the cysts are difficult to detect.
- Thorough cooking of pork kills any cysts which have escaped inspection. *Trichinellae* are destroyed if all parts of the meat reach a minimum temperature of 60°C (140°F).
- Pork kept frozen at –15°C for 20 days (or for shorter periods at lower temperatures) is also safe to eat.

Table 8.2 Characteristics of food-borne diseases

Cause of disease	*Incubation period*	*Symptoms*
Bacterial diseases *Campylobacter jejuni*	2–7 days	Fever, abdominal pain, diarrhoea.
Listeria monocytogenes	5 days to 6 weeks	Influenza-like illness, septicaemia, meningitis, miscarriages, stillbirths, infection in newborn babies.
Salmonella typhi	7–12 days	Headache and weakness, diarrhoea, fever
S. paratyphi	1–10 days	As above, but milder
Brucella spp.	1–4 weeks	Pain in muscles and joints, undulant fever.
Mycobacterium bovis	4–6 weeks	Swelling and ulceration of neck and/or intestinal lymph glands. Bone and joint inflamation.
Viral disease Hepatitis A	4 weeks	Nausea, fever, abdominal pain, jaundice.
Protozoan diseases Amoebic dysentery *Entamoeba histolytica*	up to 60 days	Diarrhoea, mild or severe. Amoebae may spread to liver and other parts of the body.
Cryptosporidium	3–7 days or longer	Acute diarrhoea
Flatworm Pork tapeworm, *Taenia solium* Beef tapeworm, *Taenia saginata*		Abdominal pain, increased appetite and wasting.
Roundworm Trichinosis *Trichinella spirialis*	1 day to several weeks	Abdominal pain, diarrhoea vomiting, fever. Later, muscular pain and soreness.

Self-assessment exercise

1 State 3 ways food-borne infections differ from food poisonings.

2 **(a)** List 5 diseases (food-borne or food poisoning) which may arise as a result of undercooking chicken.
(b) List 5 diseases which may be spread by untreated (raw) milk.

3 Which of the following organisms will multiply appreciably in 2 days when held in a refrigerator at 4°C?
(a) *Salmonella* only
(b) *Listeria* only

Sources of infection	*Foods involved*	*Preventive measures*
Cattle, pigs and poultry	Raw milk, under-cooked chicken	Use pasteurised milk. Cook chicken well.
Cattle, sheep, goats 15% of healthy people	Raw milk, soft cheeses, prepacked lettuce or cabbage based salads. Pre-cooked, chilled poultry and poultry dishes meat, pâtés	Use pasteurised milk. Cook chicken thoroughly. Pregnant women to avoid soft cheeses and meat. Keep chilled chicken dishes below 3°C for not more than 2 days. Microwave-ready dishes-heat to at least 70°C.
Polluted water Foods infected by carriers or flies.	Shellfish or watercress raised in polluted water. Food or milk infected by carriers	Chlorination of water. Maintenance of sewers. Excluding carriers from food handling, protecting food from flies
as above	as above	as above
Cattle and goats.	Raw milk	Pasteurising milk. Eradication of *Brucella* from cattle (and goats).
Cattle infected with TB.	Raw milk	Pasteurisation of milk. Eradication of TB in cattle.
Polluted water. Carriers.	Foods handled by carriers. Shellfish raised in polluted water.	Effective water & sewage treatment. Hygienic food handling. Shellfish raised in clean water.
Polluted water. Carriers. Flies contaminating foods.	Foods contaminated by infected handlers or polluted water.	Protection of water supplies. Screening food handlers for amoebic infection. Food protected from flies.
Polluted water. Infected milk.	Infected water or milk – (*Cryptosporidium* can survive pasteurisation).	Good personal hygiene. Effective water treatment.
Pigs and cattle.	Pork and pork products Beef and beef products	Effective sewage treatment. Cook meat well.
Pigs	Pork and pork products	Thorough meat inspection. Thorough cooking of meat. Holding meat frozen for 20 days.

(c) *Listeria* and *Yersinia enterocolitica*
(d) *Campylobacter* and *Salmonella*

4 List 4 precautions which should be taken to prevent Listeriosis when regenerating chicken dishes.

5 Match the incubation period and symptoms with the organism/disease.

A 2–7 days: `flu-like symptoms, abdominal pain, later diarrhoea.
B 1–4 weeks: pain in a muscles, undulant fever.
C 5 days–6 weeks: `flu-like symptoms, septicaemia, miscarriages.
D 4–6 weeks: swelling and ulcers in neck/intestines.

(a) Glandular tuberculosis. **(b)** *Campylobacter jejuni* **(c)** Brucellosis

(d) Listeriosis **(e)** Bacillary dysentery

6 **(a)** Explain what is meant by the term 'symptomless carrier'.
(b) Name 3 food-borne diseases which are commonly spread by carriers.

7 Name 3 diseases which can be spread by shellfish raised in polluted water.

8 List 3 precautionary measures which could be put into practice to prevent or reduce the spread of bacillary dysentery in a school for handicapped children.

9 Several members of the cast of *The Archers* contracted a viral disease which was traced to a carrier amongst the catering staff. They suffered from nausea, fever and abdominal pain, and jaundice. Several were severely ill and in poor health for many months. Identify the illness.

10 A friend is visiting a country where amoebic dysentery is endemic. Suggest 3 precautions he or she could take, as regards food and drink, to avoid catching the disease.

11 Which of the following pathogens is *not* affected by chlorination?

(a) *Salmonella typhi* **(b)** Dysentery bacilli
(c) Hepatitis A **(d)** *Cryptosporidium parvum*

12 Tapeworm infestation was common in children in the UK 50 years ago; it is rare today. Suggest 2 reasons for this decrease in the number of cases of the disease.

13 Describe the symptoms of Trichinosis. Give 2 precautions which can be taken to prevent the disease occurring.

14 Name 3 food-borne diseases which may be contracted as a result of swimming in sewage polluted water.

15 Name 2 diseases which have been eliminated from UK cattle by eradication schemes.

Chilled and frozen foods

Most catering establishments make use of a range of chilled and frozen foods which are convenient to use and add variety to menus.

Chilled foods

A wide range of these foods are available:

- Raw, uncooked, prepared products e.g. steaks, beef olives, chicken Kiev, hamburg steaks, fresh pasta dishes, vegetables, yeast and pastry products.
- Complete cooked dishes e.g. chicken chasseur, various beef stews and fish dishes.
- Complete meals.

When handling chilled foods you must remember that chilling is not a long term method of preservation but a means of *holding* foods fresh for a *maximum of three days*. It is essential that the foods are held at a temperature of −1°C to +2°C until required for cooking or regeneration. If the storage temperature is allowed to rise above 3°C pathogenic bacteria may become active and multiply and, in some cases, produce toxins which will not be destroyed by cooking the food at later date. Psycrophilic organisms such as *Listeria monocytogenes* are a particular danger in chilled foods as these kinds of bacteria can grow freely at low temperatures (*see* Chapter 8).

Apart from maintaining the chilled food at the correct storage temperature, care must be taken to follow *precisely* the manufacturer's instructions for preparation and regeneration. It is also essential to note the expiry dates and throw away any products which have exceeded this period.

Reheating cooked chilled foods

When reheating chilled foods in microwave ovens ensure that the correct power is being used and observe the heating time and standing time recommended by the manufacturer's of the product. If using conventional methods for reheating, again follow the manufacturer's instructions *exactly* as to temperature and time.

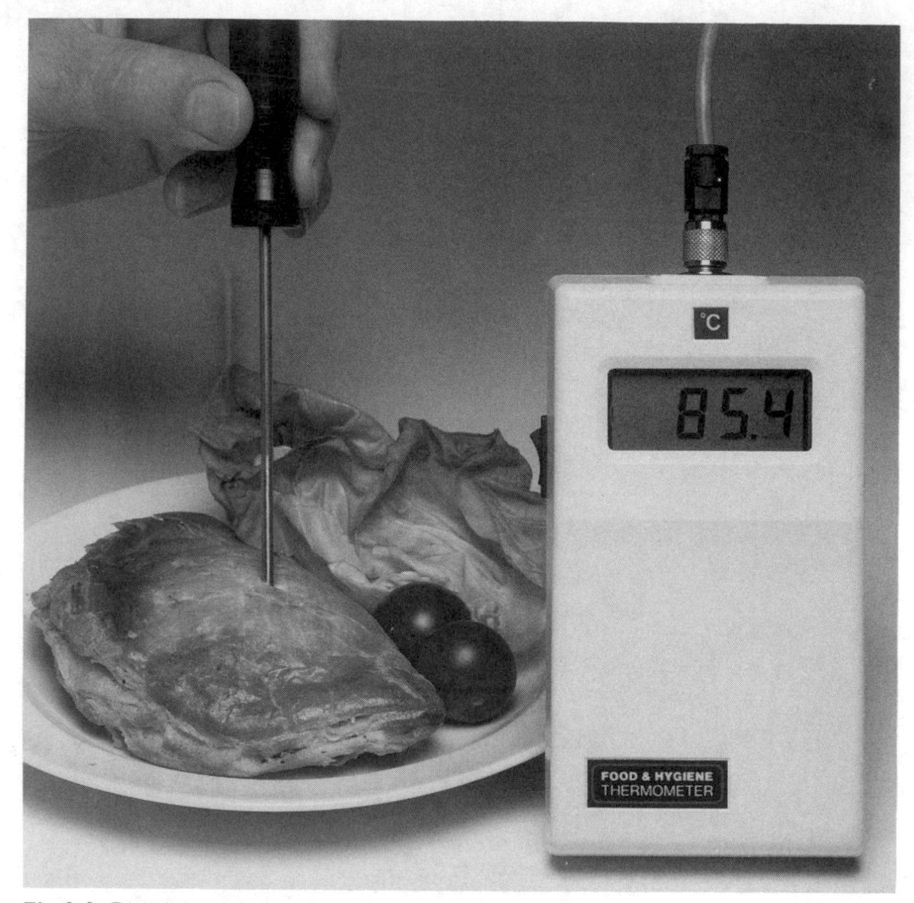

Fig 9.1 Digital catering thermomether to ensure precise temperatures are maintained.
Courtesy: Electronic Temperature Instruments Ltd

Foods vary considerably in their densities some are high – e.g. potatoes, sausages and meat products – and others are of low density, for instance sponge puddings and yeast goods. High density foods generally take longer to heat through to a safe temperature of 70°C than the low density foods.

Any chilled food which has been prepared for service and reheated or cooked must be thrown away if it is not used. Likewise, any food which has been removed from the chill cabinet in anticipation of a busy service must also be thrown away if unused. It cannot be emphasised too much that holding cooked food at chill temperatures is a hazardous operation unless *precise standards* of hygiene, time, and temperature control are followed. Manufacturers of these types of food draw up HACCP schemes to regulate:

- standards of raw materials
- hygienic methods of preparation
- precise temperatures for cooking
- precise temperatures for storing chilled foods.

It is up to caterers receiving these products to make sure that this care is continued in their own establishments as regards:

- temperature of storage
- reheating of the chilled food.

Frozen foods

Frozen foods, unlike chilled products, can be stored for prolonged periods at –18°C or below. Their storage, defrosting and cooking, however needs as much care as for chilled foods. The details have already been covered in Chapter 6.

Sous-vide cookery

The sous-vide method of cookery was developed by the French chef, Pralus. It is a development of the classical method of cooking en papillote – steam cooking in an envelope. The system involves four stages:

- vacuum packaging
- cooking en papillote
- rapid chilling
- chilled storage

This method of cooking conserves the texture, flavour and colour of the food. It prevents shrinkage in products like meat, allowing the maximum value to be obtained from expensive ingredients. When stored correctly, the products stay fresh and safe for up to six days. This means that dishes for a banquet can be prepared during the week, leaving only the last minute touches to be completed at the weekend.

Throughout the whole process strict hygiene and temperature control are *essential*. The evacuation of the pouches means that anaerobic bacteria including *C. botulinum* could multiply if there were any laxity in standards.

Stages in the system

Preparation The food to be prepared must be of the highest standard. The hygiene standards must be more akin to those expected in an operating theatre than a kitchen.

Each recipe is prepared separately on a sheet of sulphurised paper to prevent risk of cross contamination. It is then transferred with a spatula to a Cryovac pouch. These pouches are of two types :

- Shrink pouches which cling to the product when evacuated
- non-shrink pouches which are used for delicate items like fish.

Evacuation The pouches are then placed in vacuum packaging machines which slowly draw out the air and seal the product. The evacuation must remove 99 per cent of the air. The pouches are then placed in specially designed steam ovens.

Cooking Sous-vide foods are cooked in a combination oven with a steam control programme which allows cooking at oven temperatures below 100°C. The steam controller ensures that each type of food is cooked at the required temperature and for the correct length of time.

Chilling After cooking, the pouches must be cooled as *quickly* as possible. The method of chilling will vary with the volume of food prepared. Small kitchens may use iced water chillers, but for bulk operations a blast chiller will be needed.

Storage After chilling, the pouches are labelled before being loaded into a refrigerator which can hold the food accurately between 0°C and +2°C. The following items are needed on the labels to ensure the food is used in strict stock rotation.

- Dish
- Production date
- Expiry date
- Storage temperature
- Regeneration instructions

Regeneration When needed for service, the pouches are returned to the steamer for regeneration and heated until they reach 70°C.

When the correct temperature has been reached, the pouches are cut open for service to the customer (at 63°C or above).

All regenerated foods must be thrown away if not used. *They must not be rechilled or used again.*

Summary of the essential points of the sous-vide process

- Fresh, good quality foods.
- High standard of hygiene.
- Efficient vacuum packing.
- Strict control of cooking times and temperatures.
- Correct storage temperatures and times.
- Correct regeneration times and temperatures.
- Correct holding temperatures.

Cook-chill systems of food production

Many large scale institutions operate specially designed cook-chill systems. These systems require a large initial capital investment so they are only worthwhile when:

- there is a high volume of demand.
- when the demand is static – not varying from day-to-day as in many restaurants.

The system is suitable for hospitals, large firms, police canteens, school meals, welfare services and other large scale catering organisations.

Once installed, there are many advantages:

For the management Savings can be made in space as one central production unit can do the work previously done by several kitchens. Less equipment is used and it is employed more intensively. Less staff are required and those involved in the system can be employed all day. Skilled staff can be kept for food preparation instead of general duties. Economies can be made on food since the demand can be anticipated several days in advance.

For the staff The work is less rushed and they can work office hours.

For the clients The system should mean that all clients receive good quality meals whether they are near the central kitchen or distant from it. Shift workers can rely on appetizing meals instead of warmed up leftovers. Mobile regenerating units can supply emergency services with nourishing meals, day or night, in areas where no conventional kitchens are available. Even more importantly, the system cuts out the need to transport hot cooked meals, a practice which lowers the nutritional value of food and is hazardous from the hygiene point of view.

Against these advantages, there is the need to plan the whole system very precisely so that it supplies the clients' needs economically and to ensure that it satisfies the Ministry of Health guidelines for cook-chill systems.

The system

- Food is cooked in a central production kitchen. It may be cooked conventionally (though by bulk means) or by the Capkold system *(see later)*.
- It is chilled to below +3°C in a 90 minute cycle.
- The food is held in a central chilled store.
- Meals are distributed, chilled, to regeneration units.

Details of the stages of conventional cook-chill systems

1 Raw materials All food must be of assured good quality. Contracts for raw materials used in cook-chill must be drawn up very precisely and food, once received, must be stored carefully so it retains its freshness.

2 Preparation Raw materials must be prepared in separate areas from the cooking areas. Separate machinery, utensils and knives must be dedicated for raw and cooked food so there is no possibility of cross contamination. Joints of meat must not exceed 2.5kg in weight.

3 Cooking The cooking process must ensure that all vegetative pathogens are killed. Food must be held at an internal temperature of 70°C for at least 2 minutes to kill *Listeria monocytogenes*. A probe thermometer must be used to check that this temperature is reached in the densest part of the food.

Review exercise

Why is it particularly important to ensure *Listeria* are killed in a food which is to be held at chilling temperatures?

Portioning If the food is to be portioned, this process must take place as soon as possible after cooking – within a maximum of 30 minutes after cooking. The portioning should take place in cool room with a temperature of 10°C. It is very important to have clean containers of stainless steel, aluminium or porcelain, or good quality disposables. To assist the chilling process, the food should be spread evenly and to no more than 50mm depth.

If reusable trays are employed, the washing equipment must be sited *away* from the food handling areas.

Chilling Chilling must take place *immediately* after portioning. The food is placed in specially designed rapid chillers which are capable of reducing the temperature of the food from 70°C to +3°C or below within 90 minutes or less when fully loaded. The chiller must have automatic controls, a thermometer accurate to +/– 0.5°C, and a temperature chart recorder.

Chill store The refrigerated store used for holding the chilled foods must be specially designed for the purpose and *only used for products from the cook-chill process*. It must maintain the foods within the temperature range 0°C to +3°C. The temperature of the store must be displayed on an accurate recording device. Alarms should be fitted to the chill store which will alert the switch board or security staff if the temperature of the store rises above the critical temperature. An increase to 5°C may be permitted for *very short times* – for instance when the chill store is on defrost cycle.

Should the food reach +5°C during storage and distribution but before regeneration, the food must be used as soon as possible – certainly within 12 hours – or else the food must be thrown away. If it reaches +10°C it should not be used and should be discarded immediately.

All products must be labelled as for sous-vide foods and strict stock control must be enforced. Providing the strict requirements of the system have been met, the food has a maximum storage life of 5 days *including* both the *day of production* and the *day of consumption*.

Distribution of chilled food

Chilled food must be maintain at or below +3°C during distribution prior to regeneration. When the distance between the central production unit and the satellite kitchen is short, distribution may be by means of insulated containers which are chilled before use. When the food has to travel longer distances, a refrigerated vehicle is needed. The temperature of the food must be checked on arrival at its destination and, if it is +3°C or below, it must be placed in chill storage until required for regeneration.

In hospitals, the distribution is usually done in specially designed trolleys which maintain the food at chill temperatures during transport and are then plugged in to the electric supply on the ward to raise the temperature of the food, rapidly to 70°C. This system requires good organisation and vigilance to work effectively. The trolleys need checking at frequent intervals to make sure that they are holding the chill temperature effectively and to ensure the food is being heated to 70°C as quickly as possible. It also requires careful organisation to ensure that service of the food can start within 15 minutes of the food reaching 70° and that it does not drop below 63°C before it reaches the patient. This not always easy to achieve on a busy ward.

Regeneration

Apart from the hospital system mentioned above, chilled food is regenerated in a satellite kitchen near to the point of service. Foods are regenerated as soon as possible after withdrawal from storage in infra red, microwave or convection ovens.

How safe is cook-chill?

Each individual system has to be considered on its merits. If the system is operated as intended – with strict temperature controls, rigorous hygiene standards, and regeneration near to the point of service – it should be safer than conventional methods of bulk catering. However, there has been concern about *some* cook-chill systems in hospitals because:

- In many cases existing buildings have had to be modified to operate the system and the capital costs have proved to be too great. This has resulted in compromises being made in the system.

- A considerable number of hospitals do not follow the Guidelines and reheat chilled food *centrally* and distribute the food 'hot' to the wards. This cancels out the benefits of the system as regards the safety and nutritional value of the food.
- The distances between central kitchens and wards can be considerable and the difficulties of organising service of food soon after regeneration can be greater than in other catering situations.
- Patients are sick, elderly or very young and so particularly vulnerable to any pathogens in their food.

Capkold is one alternative to the conventional cook-chill system which may solve some of the problems of putting chill systems into hospitals and welfare catering. In this system the foods are vacuum packed as in sous-vide and steam cooked and later cooled in the same Cooktank or kettle.

The Capkold system

The foods are vacuum packed in cryovac cases, clipped and tagged and then cooked in the kettle by steam heated water at 70°C to 85°C. A probe thermometer in the centre of large joints ensures that they reach the correct end temperature. When the cooking cycle has been completed, chilled water is circulated until the food reaches +2°C (on average in 75 minutes). The whole process is automatically registered on a chart recorder.

The Cooktank can also be used for fluids such as sauces, gravies and minced meat. After cooking, these products are pumped into cryovac pouches and tumble chilled in less than 60 minutes to below +3°C. All cryovac foods can be held at +2°C for up to 21 days.

For regeneration, the packs are opened and emptied into gastronorm dishes and heated for quick service.

The Capkold system has a number of advantages over the conventional cook-chill system:

Safety

- Food is not exposed until unpacked in regeneration kitchens so there is no chance of contamination during portioning or in transit.
- The whole operating system is automatically controlled and temperatures are accurately regulated.

Economy

- The slow cooking method gives high yields with expensive foods like meats.
- Labour costs of the cooking/chilling stage are minimal. Two men can run the whole process.
- The units are compact and only require hot and chilled water to service them.

For hospitals, these savings on the cooking and chilling stages could be used for building and staffing smaller satellite regeneration units near two or three wards, cutting out the inconvenient and potentially hazardous transportation systems.

Cook-freeze

Cook-freeze is another system involving advance preparation and cold storage of food for widespread distribution. Because the storage temperature is much lower, the products have a much longer shelf life than those of cook-chill. In general, precooked frozen foods can be stored in good condition for eight weeks and in some cases for much longer periods. The system is used in airline catering and other bulk catering systems.

All the rules and procedures relevant to cook-chill apply to the cook-freeze system, with the exception of the temperature and storage times.

After cooking and portioning, foods are blast frozen. They must reach –5°C within 90 minutes and be further deduced to –20°C before storage. A fast rate of freezing is *essential* to prevent large ice crystals forming which would damage the structure of the food. The food must be kept at –20°C until required for regeneration.

Certain foods are adversely affected by freezing and thawing. These are foods such as sauces and gravies which are egg or starch thickened, emulsion based products such as salad dressings, and sweets where the structure is due to egg foams. These types of products have to be specially formulated to freeze successfully.

Refrigeration equipment

Chillers Chillers are designed to reduce the temperature of raw and cooked foods to 1°–2°C in a maximum of 90 minutes. Cold air is circulated around the chiller by a fan. Chillers must *never* be loaded beyond the maximum capacity recommended by the manufacturer. Too much food in the chiller will result in a longer chilling time, which will increase the danger of food poisoning and may cause deterioration in the food. Chillers must not be used for storage of chilled food. When processed, the chilled food must be transferred to a refrigerator to be held at 1°C–3°C.

The chiller must be defrosted periodically so that ice does not build up on the cooling unit. The internal temperature must be checked regularly to make sure it is holding the correct temperature. The unit must be washed and cleaned at least once a week. Modern chillers have automatic defrosting and temperature recorders fitted.

Blast freezers These are designed to reduce the internal temperatures of foods to –20°C or below as rapidly as possible. They work on the same principle as the rapid chillers and require the same care and maintenance.

Refrigerators These cabinets can now be purchased with pre-set temperatures for specific types of foods e.g. at 1°C–3°C for chilled foods. Modern freezers, blast chillers and deep freezers are fitted with external thermometers and alarms. These may be of the dial pattern showing temperatures of 1°C and above in red and –1°C and below in blue. If the temperature rises above the safety level a buzzer sounds or a light flashes to alert staff of the fault. Temperatures of such equipment should be checked weekly and any rise in temperature recorded and reported to the refrigeration engineer.

Regeneration Regeneration or reheating of chilled or frozen foods can be carried out by using any of the following equipment:

- microwave ovens
- combi-ovens
- micro-aire ovens
- infra red ovens
- steaming and regothermic ovens.

The oven must be capable of raising the internal temperatures of the food very rapidly to 70°C.

Self assessment questions

1 List 3 reasons why particular care must be taken in storing chilled products.

2 Strict temperature control is necessary in sous-vide cookery to prevent growth of the deadly anaerobic bacterium ______ ______.

3 List the items of information required on the label of a chilled food.

4 If a cook-chill product is processed on a Monday, which is the latest day it may be regenerated and served to the client?

5 List the advantages of the Capkold system over the conventional methods of cook-chill.

6 Give the correct figures for the following:

(a) The minimum time and temperature to ensure destruction of *Listeria* bacteria in food prepared for cook-chill.
(b) The maximum time interval allowed between cooking and portioning in conventional cook-chill systems.
(c) The temperature a food must reach at the end of a rapid chill cycle.
(d) The temperature range for storage of sous-vide products.
(e) Chilled and frozen foods must be regenerated to a temperature of _ _°C and served at no lower than _ _°C.

7 Briefly list the advantages of installing a cook-chill or cook-freeze system at an establishment catering for large numbers of people.

Task

The guidelines on cook-chill and cook-freeze recommend personnel operating these systems to adopt a HACCP approach.

Draw up HACCP flow charts for the sous-vide and cook-freeze methods of food production.

10 Alternative forms of catering

Traditionally people went out to hotels and restaurants to dine; the work force ate in canteens or self-service restaurants and students in halls or refectories. Today, the caterer often comes to the customer – offering take-away food to eat at home, telephone ordering and delivery services, the organisation of parties or functions at home, etc. Similar services are offered at public events such as shows and sporting events, while roadside mobile cafes and trailers sell hot dogs and beefburgers to the motoring public.

All these alternative catering outlets, whether large or small, are food businesses within the meaning of the Food Hygiene Regulations, but because they are so many and various it is often difficult for the authorities to regulate their standards. Some are run by internationally known organisations which regulate the quality of raw materials, food handling and service standards and the training of staff. Others are more haphazard and are carried out in cramped and often unsuitable conditions.

This chapter deals with a number of types of alternative catering outlets, the problems encountered, and the standards they should maintain to safeguard the public.

Mobile food outlets

These outlets have increased dramatically over the last few years and are to be seen on most major roads throughout the country. They may be housed in caravans, converted coaches or buses, vans or purpose built trailers and units. The foods offered for sale include cooked breakfasts, hot dogs, hamburgers, sandwiches, light snacks and a range of beverages and soft drinks. To offer this range of foods, the vehicles need to be equipped with a refrigerator for storing high risk foods such as cream, milk, butter, cold meats and any fillings for sandwiches. The refrigerator should be capable of holding the foods at 4°C. These outlets also need a small stove for heating hot dogs, frying beefburgers, bacon, eggs and tomatoes. You can imagine for yourself the high temperature which will be produced by cooking in such cramped surroundings and the danger if food is not stored correctly.

A *very high* standard of hygiene is necessary to prevent cross contamination occurring in such a confined working area. Very few of these units are fitted with a deep freeze, so the refrigerator has to be used for raw meat, cooked meat and dairy products. This means that extreme care is necessary in arranging the foods to prevent cross contamination. Staff hygiene is also very important – hands must be washed thoroughly following each time raw meat is handled. There should be two washing areas, one for hand washing and one for crockery and cooking utensils. The hand washbowl should have soap, nail brush and a clean towel or paper towels, and there must be sufficient hot water and detergent for washing equipment.

All hot dogs, sausage rolls, meat pies and beef burgers must be cooked or reheated to reach an internal temperature of 70°C or above and be held at least 63°C for sale. Any food not sold within an hour of heating must be thrown away.

Sandwiches and fillings must be prepared on clean surfaces which have been washed with a sterilant. The sandwiches must be wrapped and kept at 6°–8°C. At the end of the day all unsold sandwiches must be thrown away.

All persons operating these outlets must be dressed correctly – preferably in a white coat and apron and suitable hat to cover the head. A first aid box must be available in the van so injuries can be treated.

At the end of each day, the vehicle must be cleaned *thoroughly*. All refuse must be removed, never left in it overnight. All the cooking equipment must be cleaned, and spillage and food particles removed. The preparation surfaces must be washed and sanitised and the floor scrubbed to remove all food particles and grease. The refrigerator must be cleaned and any unsold defrosted meat products and unused sandwich fillings, thrown away. When locked up for the night, there should be some form of ventilation to allow fresh air to circulate through the vehicle.

Public house catering

Nearly all public houses serve hot and cold foods, either as bar meals or snacks, or full meals in their restaurants. Some establishments buy in prepared foods, while others prepare their own food on the premises. The range of foods offered as bar meals usually includes chilli-con-carne, fried scampi, fried fish, chicken, steak pies, cottage or shepherd's pie and lasagne. These dishes with Cornish pasties, assorted sandwiches, ploughman's lunches, a variety of salads, and sweets complete the menu of most public houses.

Most of the hot dishes are pre-cooked and frozen and then heated by microwave ovens. This method of operation requires careful control during preparation, cooking, cooling and storage to ensure safety.

Storage of foods

All the food products for preparation of cooked and uncooked dishes must be of good quality and fresh.

Foods should be stored at the temperatures given below. It is preferable to have separate refrigerated units for the different categories.

- Raw meat and fish 2°C–4°C
- Cooked meat and fish 2°C–4°C
- High risk foods, cream, meat pates, soft cheeses and milk 2°C–4°C
- Raw vegetables and salad items 4°C–6°C

In small establishments which have only one refrigerator, the foods must be stored as follows at temperatures of 2°C–4°C.

- Top shelf – high risk foods
- Middle shelf – cooked meats and fish on separate trays or plates covered with cling film.
- Lower shelf – raw vegetables and salads on trays, properly covered.
- Bottom shelf – raw meat and poultry covered with cling film.

Frozen foods must be held at –20°C until required for cooking or defrosting. Frozen sweets and gateaux must be defrosted in a cool place, at 10°C. Ice cream sweets must remain at –20°C until required.

Preparation

- All foods must be prepared on sanitized work surfaces. The working area in public houses is usually quite small so it is essential to have good working practices and *clear as you go*.
- All prepared foods not required for immediate cooking must be refrigerated.
- Food preparation staff should be correctly dressed i.e. white coat, white apron and the head suitably covered.

Cooking

- Foods must be cooked so that they reach an internal temperature of 70°C–85°C.
- Foods which are to be kept hot must be maintained at 63°C or above.
- Foods which are to be cooked and then frozen **must** not be more than 50mm (2in) deep and cooled to 2°C as quickly as possible but within a *maximum* time of $1\frac{1}{2}$ hours. They **must** then be placed in the freezer to reduce the temperature to –20°C within $1\frac{1}{2}$ hours.
- Foods which are to be cooked and chilled must be used within 3 days of preparation.
- Cooked and chilled foods **must** be cooled quickly to 3°C, covered, dated with the production and expiry date and stored at 1°C–3°C. The refrigeration unit must maintain this range of temperatures and *must not be used for storing other foods*.
- Any chilled food past the expiry date must be thrown away.
- Any food which has been kept hot or removed from the chiller in anticipation of being required and not used *must be thrown away.*

Reheating chilled frozen foods

The microwave oven used for reheating these foods must be an industrial model capable of reheating the food to the required temperature as quickly as possible. The operator must be fully conversant with the operation of the oven and in reheating dishes.

Reheated food **must**:

- reach an internal temperature of 70°C and be held at that temperature for at least 2 minutes,
- be served when it is ready or held at or above 63°C until required – but no longer than $1\frac{1}{2}$ hours.

The staff responsible for the production of chilled and frozen dishes must be fully aware of the danger of food poisoning that would arise if the stated procedures were not carried out correctly. Thermometers for checking the foods are available and easy to use (*see* p 134).

Preparation of cold foods

All foods to be served cold must be kept cold before and after preparation and not left in a warm kitchen or room until required. Personal and food hygiene are *particularly important* when preparing cold foods as they are *not* cooked or reheated before being served.

Sandwiches and filled rolls

If these are prepared in advance, they must be individually wrapped or covered with cling film and kept refrigerated at 6°C–8°C. If not sold at the end of the day (12 hours), they must be thrown away. The fillings must be kept at 2°C–4°C. This is particularly important in the case of cooked meats, meat pâtés, egg mixture with mayonnaise, prawns, crab, tuna and salmon.

If sandwiches are made to order, the work surface must be sanitized and kept *only* for making sandwiches. The fillings must be kept refrigerated until required and replaced after use. Any knives used must be washed and dried after each order.

Salads

Prepared salads for a salad bar must be kept refrigerated before being placed on display or kept on a bed of ice. The salads **must**:

- be protected by a sneeze guard
- have separate cutlery for each salad
- be thrown away at the end of each service i.e. lunch and evening.

If a refrigerated display is not available, then small quantities must be displayed and replenished as required from stock kept in the refrigerator. If a sneeze guard is not fitted, the salads must be kept covered.

Ploughman's lunches

This is a very popular food item in public houses and the variations range from the traditional cheese to cold meats, pâtés and varieties of fish. There may be a choice of cheeses offered to the customer – if so it is permissible to have *small* quantities of cheeses at room temperature, providing they are covered. This is to restore the flavour. If taken straight from the refrigerator, the flavour is not at its best. Meats, pâtés and fish items *must* be kept refrigerated at 2°C–4°C.

Hot display cabinets

If the establishment has a hot display cabinet for sausage rolls, Cornish pasties and meat pies, then these items must be heated to 70°C for at least 2 minutes before they are placed in the cabinet which must maintain them at, or above, 63°C. *The food must not be heated in the display cabinet* because this would not bring the food to the required temperature.

Any food left unsold after the lunch and evening hours must be thrown away.

Speciality eating houses/take-aways

The number of different speciality and ethnic restaurants and take-aways has increased markedly in the UK. These range from spit-roast chicken houses, Turkish, Italian, Chinese and Japanese restaurants, fried chicken and hamburger take-aways and restaurants. The standard of food and personal hygiene vary from one take-away to another and this presents a problem of food safety for the consumer.

Chicken take-away houses

As with all chicken dishes, the cooking method must ensure that *Salmonellae* are killed. This needs care when spit roasters are used as they are not as efficient in raising the temperature of the meat to the required 70+°C as conventional ovens. It is essential to allow sufficient time for all joints to reach the required temperature and then to hold them in a hot cupboard at, or above, 63°C. The chicken must not be left on the spit with the heat turned off. The joints must be held at 63°C or cooled rapidly and sold cold.

Fresh chicken must be kept refrigerated at 2°C–4°C until required for cooking. Frozen chickens must be *thoroughly* defrosted in a refrigerator at 10°C over a period of 24 hours before being spit roasted.

Fried chicken take-aways

These establishments use commercially frozen chicken portions. The portions when deep fried must reach a temperature of 70°C–85°C to ensure thorough cooking, otherwise there is a *high* risk of food poisoning. Deep fryers must be checked periodically to ensure the correct temperature is registered. The fryers must not be over filled as this will reduce the temperature of the oil – thereby extending the cooking time. For example, ten chicken pieces placed in the fryer at 180°C should take 7 minutes to cook. If 15 pieces are added to the oil and cooked for 7 minutes they will not be properly cooked because the temperature of the oil will have been too low.

The chicken pieces must be kept at –20°C until used. If chilled pieces are used instead, they must be stored at 1°C–3°C and thrown away after the expiry date. *All* unsold cooked chicken pieces must be thrown away.

Kebab houses

Doner kebab houses are to be found in many cities throughout the UK. The kebab is a large joint (often 45kg) of minced lamb or beef which is hand moulded onto the spit on which it is cooked. The spit rotates slowly in front of a vertical radiant grill. The meat from the surface of the kebab is sliced off and placed inside Greek bread to be sold to the public. This method of food preparation involves a series of steps which are particularly hazardous as regards food hygiene:

- The kebab is composed of *minced meat*, not one solid joint of meat. Mince meat is known to have a much larger load of bacteria than is found in the equivalent weight of solid meat, since the mincing of meat creates a large surface area for bacteria to grow upon. The bacteria which were once only on the *surface* of the meat are spread *throughout* the mince and provided with a convenient source of food and moisture in the juices expressed from the meat.
- The kebab is *moulded by hand*. This means that there is a good chance of exchange of bacteria between the meat and the hands of the staff preparing the kebab. Unless extreme care is taken, *Salmonellae* are likely to be spread from the hands to other foods and the environment generally.
- The kebabs are usually *large* and may be difficult to refrigerate without contaminating other food or equipment.
- It requires great skill to cut only the layer of meat which has been *thoroughly* cooked without contaminating the knife with juices from the undercooked meat beneath.
- As in other types of catering, there are busy times and slack periods in kebab houses and this can lead to dangerous practices such as cutting off a large number of slices of meat when a rush of customers is expected and keeping them warm in a bain marie.

- Another dangerous practice is switching off the heat on the grill when the demand is low. This allows the temperature of the meat to drop to a level at which bacteria can grow.

Some modifications can reduce the dangers inherent in doner kebab cooking:

- The *fresh* mince should be refrigerated until required.
- Equipment and hands should be washed thoroughly *before and after* mixing and moulding the kebabs.
- Making several smaller kebabs is much safer than one large one.
- The kebabs should be refrigerated until set.
- Once cooking has commenced *the heat should not be turned off* in slack periods.
- Only *thin layers* should be sliced off from the outer layer when *thoroughly* cooked.
- The knife used for carving should be *washed thoroughly* between slicings.
- The meat must be served *at once* and not held warm.

Asian and Far Eastern take-aways

These establishments offer a wide range of dishes with a rice accompaniment. Most of the food is pre-cooked and reheated to order for the customer. Because of the number of dishes offered, the items of food required in readiness are considerable – a variety of fish, shellfish, eggs, pork, lamb, chicken, fruit, vegetables and many speciality foods.

If the system is to be operated safely, the establishments need:

- Freezers to keep the meats and seafoods frozen at –20°C until required.
- Cool areas for cooling the prepared foods rapidly to 3°C prior to refrigeration.
- Sufficient refrigerated units to store the chilled food safely until needed for reheating.

The meats and shellfish need careful defrosting so they do not cause cross contamination. Failure to defrost items like shellfish completely before cooking is particularly hazardous.

The rice which accompanies the dishes may be boiled in saucepans or cooked in automatic rice cookers which switch off when the rice is cooked leaving it to stay hot in the insulated container. Rice as we saw earlier, is frequently infected with *B. cereus* spores.To prevent the spores germinating and causing food poisoning, the rice must either:

- be fried or used immediately
- be kept at or above 63°C until needed
- be properly refreshed, (*see* p 96) and refrigerated until required.

Meat stock made by boiling bones is used in preparing many of these dishes. It is essential that the stock is kept near to boiling point all day and a *fresh* batch made the following day.

At the end of the day all foods which have not been kept refrigerated must be thrown away.

Pizza houses

Pizzas are less likely to cause food poisoning because the fillings are much thinner than for the other foods described and because the dough base is cooked at a high temperature so that it reaches 70°C–85°C quickly. However, the wide range of fillings that are used require correct storage – particularly the fish and meat toppings. Meat should be stored at 2°C-4° and fish at 1°C–3°C. In busy establishments small quantities of the filling may be at hand and replenished from the refrigerator as required. Any products left unrefrigerated at the close of business *must* be thrown away.

Delivery services

Establishments which operate phone order and delivery services must ensure that the following rules are observed:

- The food is properly cooked to an internal temperature of 70°C–85°C
- All food containers are clean and completely sealed,
- Filled containers are placed in insulated boxes which will maintain the temperature of the food at a minimum temperature of 63°C for at least 30 minutes, to allow for delivery to the client. Polystyrene, waxed or tinfoil containers are not suitable because they are not sufficient to maintain the temperature to 63°C.
- The insulated containers and the delivery vehicles must be kept clean.

Hygiene

A high standard of hygiene is essential in all areas of take-away food preparation.

- Staff should be properly dressed, with clean white overalls, aprons and head gear.
- All working surfaces, tables, equipment and counters must be kept clean and sanitized throughout the day.
- All equipment must be in good working order and refrigerators checked regularly for the correct temperatures.
- Premises must be thoroughly cleaned at the end of each day and rubbish removed from the kitchen and service area.
- All sauce bottles and condiment containers must be kept clean.

Outdoor catering functions

Outdoor catering firms are employed at large sporting occasions such as Ascot and Henley, motoring events, and at agricultural and air shows. The services offered range from the small caterer operating from a mobile vehicle to large operators serving thousands of meals, both hot and cold. The food outlets may be spread all over the show in separate marquees. Each marquee will have its own preparation kitchen which is supplied with food from a central kitchen or storage centre. The food may be cooked on site or at the company depot and transported in refrigerated vans. Most of these events take place in summer. After reading this chapter you will realise the dangers inherent in these situations – where large quantities of cold foods are handled with the minimum of refrigeration.

The facilities on site on these occasions may be very poor, so everything has to be brought in – even water, by tanker. The operation and organisation of these events is beyond the scope of this book but the extra care required as regards food hygiene in these situations is discussed.

Central kitchen storage centre

Mobile refrigeration units should be provided in these areas with separate units for raw fish and meat, cooked meat and fish, dairy products and for vegetables, salad items and fruit, working at the temperatures given on p 155. All fresh deliveries of food must be labelled with delivery and expiry date and used in date order. Any food found to be past the expiry date must be thrown away. The refrigeration units must be kept clean and not overstocked, otherwise the required temperature will not be maintained. All foods must be properly wrapped and transported under refrigeration.

Service kitchens

The service kitchens attached to the marquees where the meals are served are equipped with preparation tables. These must be covered with a suitable easy-to-clean coverings e.g. white industrial polythene sheeting which must be sanitized and renewed when necessary. A water boiler is *essential* for providing a constant supply of hot water for washing hands and equipment. The wash-up area must be situated *away* from the main preparation area. The equipment wash area must have a supply of suitable detergent and baths for washing and rinsing. The hand wash area must have soap, nail brush and a supply of paper towels. Other equipment may include cooking stoves, ovens and hot cupboards. The hot cupboards must be capable of keeping food hot at 63°C and *must be checked* after being connected to the bottled gas supply.

The foods must be distributed to the service kitchen as late as possible, allowing time for preparation. This reduces the time foods are without refrigeration before being consumed.

During the preparation of the food, staff must work methodically and hygienically and *clean as they go*. Any food left over after the function must be thrown away and not saved for future use.

Unskilled labour is often recruited to carry out mundane tasks and to plate up food. These people must be instructed in the importance of good hygienic practices. The manager or chef in charge of the unit must enforce these rules:

- Handle food as little as possible.
- Keep all foods covered before and after preparation.
- Wipe down all surfaces frequently with a sterilant.
- Wash your hands every time you handle containers.
- Do not smoke in any food preparation area.
- Refuse and food waste must not be allowed to accumulate in the food preparation areas. It must be removed to the collection bins.
- At the end of the day, scrub and santitize the whole area and throw away any left over food.

You will now realise the difficulties managers and staff encounter in maintaining the high standards of hygiene and working methods which are essential at functions of this type. Environmental Health Officers are aware of the difficulties and frequently make advisory visits or spot checks to functions taking place in their areas.

11 Storage of food

The food storage area should be situated on the north side of the building as this will be naturally cooler. The store must be well ventilated, dry, vermin-proof, and easy to keep clean. The necessary fixtures and fittings include:

- Storage cupboards and ample shelves for stacking foods
- Large bins on castors for sacks of flour, rice etc.
- Stainless steel tables, food scales and a selection of knives for cutting foods
- Suitable tools for opening packing cases
- First aid box
- Wash hand basin, properly equipped
- Step ladders for reaching the higher shelves.

The lower shelves of the food store must be at least 450mm (18in) from the floor.

Dry goods area

This area is used to store foods such as:

- flour, pulses, dried fruit and pastas
- prepared mixes for soups, cakes and sponges
- canned and bottled food e.g. tinned fruits, vegetable preserves, olives, gherkins, bottled sausages and shellfish.

It is important to stack these foods properly on the shelves to prevent them falling off or becoming damaged. The heavy packs should be placed on the lower shelves and the lighter ones higher up. Items which are used extensively should be stacked at or below waist level; those rarely used, on the higher shelves. When refilling the shelves, the old stock must be brought forward and new stock placed behind so they are used in the correct order (stock rotation). Any food past the expiry date must be removed and thrown away. Some foods lose their colour if placed in strong light, e.g. ground spices, bottled products dried herbs, etc. These must be stored in dark cupboards.

Points to look for when stacking or checking dry stores

Dried foods Infestation by various weevils will be indicated by a fine powdery deposit or a 'webby' appearance. Weevils are visible to the naked eye as tiny brown–black specks. Throw away all affected products (*see also* Chapter 13).

Dried fruits If they have become moist, they will give off a fruity aroma of fermentation. Throw away any affected fruit.

Canned foods Badly dented cans may show signs of rusting or seepage which indicate that they are no longer air tight. Blown cans show signs of bulging at the top or bottom. Any cans with these faults *must not be used* – return them to the supplier or throw them away.

Bottled foods These may show signs of leakage from around the lids indicating that the seal is faulty and they are no longer air tight. Return to supplier or throw them away. *Do not use them.*

Weighing foods for orders

When weighing foods for the kitchen always:

- use a clean scale pan
- use clean knives and spoons
- place the foods in clean containers, suitable for the purpose
- keep raw, unprepared foods away from prepared and cooked foods
- wash your hands thoroughly after handling any raw foods before you handle cooked foods
- wash your hands thoroughly after opening packing cases or stacking foods. You will be surprised how dirty your hands become after these jobs.
- Wash down and sanitize all table surfaces and chopping boards when you have finished. Also wash and sterilise knives and other equipment.

Note: You must never smoke in a dry foods area or any area of food storage, or where there is open or uncovered food.

Refrigerated section

Adjacent to the dry storage area there should be a refrigerated section for the storage of perishable goods. Each category of food must have a separate refrigerated unit running at the correct storage temperature:

• meat and poultry	2°C–4°C
• meat that is hung	1°C–3°C
• fresh fish (with wet ice containers)	1°C–3°C
• cooked meats, including smoked meats	2°C–4°C
• dairy products	2°C–4°C
• vegetables and fruits (not bananas)	4°C–6°C
• deep freezer	–20°C

All refrigeration units must be cleaned regularly and temperature checks carried out (*see* Chapter 9). The floors of walk-in units must be cleaned when necessary, at least twice a day, depending on use.

If any raw meat or poultry is to be cut up before distribution to the larder, it must be cut on a table in the refrigerated section, kept for this purpose. It *must not* be cut up in the dry store section. This will prevent any cross infection taking place from the meat to dry or ready to eat foods. Hands must always be thoroughly washed after handling raw meat, poultry and fish, before doing anything else.

Goods inwards delivery point

This point must be situated as near as possible to the storage area. Ideally, a raised platform should be provided to assist the unloading of supplies with the minimum of lifting. It allows trolleys to be wheeled into the delivery vehicle, loaded and taken directly to the storage area. This is essential for delivery of frozen foods, raw fish, meat and other perishable goods so that the time they are without refrigeration is cut to the minimum. The food can be checked as it is loaded onto the trolleys and weighed, if required, before it is stored. Food must never be placed directly on the ground whether wrapped or unwrapped. The dirt picked up by the wrappers or cases can be transferred to unwrapped food.

All fish and poultry should be delivered by refrigerated vehicles in suitable clean containers, *properly covered*. If suitable containers are not available clean polythene bags may be used.

The delivery point area must be kept clean at all times and regularly scrubbed. In dry weather, a sprinkling of water should be used to lay the dust and stop it blowing into the storage are.

Rubbish must never be allowed to accumulate. Empty crates, baskets and trays belonging to suppliers must be washed and stored in an appropriate place to be collected. The steps of the delivery point must be kept in a good state of repair and free from any obstruction.

Trolleys

The trolleys used for stores and distribution to the kitchen must be colour coded and used for the stated purpose only, e.g.:

- raw meat — red
- raw fish — yellow
- vegetables/fruit — blue
- dry store — green
- dairy produce — black

The trolleys must be properly maintained, with all wheels running smoothly to ensure smooth running when loaded. After each delivery, the trolleys must be steam cleaned and kept in an appropriate place. Those used for meat and fish must be steam cleaned after every use.

Storage life of frozen products

When storing away fresh deliveries of meat, poultry and fish, you must remember to bring the old stock forward to use first and place the new stock behind. It is particularly important with these products as they may not be marked with an expiry date. Frozen foods must also be stacked away so the old stock is used first. They do not have an expiry date but have a maximum recommended storage life when kept at –20°C.

- Dairy products 1 month
- Beef, lamb 10–12 months
- Pork 3–6 months
- Sausages 1 month
- Vegetables/fruits 10–12 months
- Fish 3–6 months
- Shellfish 3–4 months
- Poultry 8–10 months

Cleaning of stores areas

Cleaning schedules should be drawn up to ensure regular and thorough cleaning. The tasks that need to be carried out are:

- Removal of all food scraps
- Cleaning containers, inside and out
- Floor areas must be scrubbed
- Shelves and cupboards cleaned
- Scales and pans washed
- All walls and ceilings washed
- All table surfaces, legs and under shelves to be washed and sanitized
- Wash hand basin to be cleaned
- Refrigeration units to be cleaned
- Delivery point cleaned

You can prepare a cleaning schedule from the tasks listed to ensure cleaning of the whole area over a two week period. If you need any assistance, turn to pages 174-5 in Chapter 12.

Maintenance

- The floor must be kept in good repair, any broken or loose tiles should be replaced or reset.
- The shelves must be securely fixed and kept in good order.
- The walls and ceiling must be kept in a good state of repair. Any flaking paint must be removed and the surface redecorated.
- Sliding doors to cupboards must be attended to, so that they slide smoothly.

Safety in the stores

Using step ladders

- When using steps to reach the highest shelves, make sure they are firmly on the ground and open to their fullest extent
- Use steps which give you easy access to the items required.
- Do not over stretch to reach items, descend and move ladder nearer to what you require.
- Make sure the steps and your shoes are free from grease.
- When not in use, keep the steps in a safe place, preferably on a hook against the wall in a corner.

Lifting heavy items

- Do not attempt to lift or carry items which you cannot carry with ease.
- Always ask for assistance. You are not admitting you are weak, just being sensible. You could end up with permanent back problems if you try lifting something too heavy for you.
- When lifting heavy items, bend the knees slightly, keep your back straight and take hold of the load. Push up, straightening the legs and keeping the back *straight*.
- When giving or receiving assistance to lift or carry heavy items, make sure the other person is ready and then lift or move together.
- Do not mess about when lifting with another person – it is potentially dangerous.
- Do not drop the items as this could damage the contents or split the casing. Bend the knees and lower one end to the ground first.

Stacking cases and sacks

- Never stack cases or sacks directly on the floor – always use duck boards or pallets. This prevents the cases from getting wet when the floor is washed and also allows a circulation of air around the stack.
- When stacking cases of the same size be aware of the height of the stack – remember the cases have to be taken down at a later date.
- Stack smaller cases on larger ones. Putting the larger ones on top makes an unstable stack which could fall, spoiling the contents or causing injury to staff.
- Always stack sensibly and never too high.

12 Cleaning

In previous chapters we have seen how important it is to keep food premises scrupulously clean. Cleaning involves the removal of 'soil', the unwanted material which contaminates the kitchen environment. Soil, however, is not one simple substance but a complex mixture of materials. Some of it is dust, formed through friction constantly wearing down solid surfaces. The air blowing in from outside premises brings in inorganic materials such as brick and stone dust from buildings, as well as ash and soot from the combustion of fuels. In addition, peoples' footwear carries in earth and other materials from the street outside.

There is also a surprising amount of organic material – pollen from plants, animal and human hair, dust from bird feathers and, most of all, scales of human skin and the bacteria which find lodging there.

Loose particulate dust on its own is easy to remove by dusting or vacuum cleaning, but in a kitchen it is always accompanied by other materials which are much more difficult to lift from surfaces. The main problem soils are discussed below.

Grease is the main problem in kitchens. Cooking releases a constant stream of hot greasy vapours into the air which later condense on any cold surface and make the removal of soil much more difficult than it would be otherwise.

Hard chalky deposits often form in urns, kettles, and drinks dispensers. This happens because much of the water we use for cooking and cleaning is 'hard'. This kind of water contains calcium or magnesium salts which deposit as scale when the water is heated to above 60°C and is then cooled. The scaly deposits dull crockery and cutlery as well as coating the inside of boilers and urns, making them less efficient when heating water.

Milkstone is a similar substance formed from the calcium and magnesium phosphate and protein in milk. It produces a scaly deposit in drink dispensers and pans used to boil milk.

Dripping taps often leave persistent coloured **stains** on sinks and baths. They are caused by oxidation of the metal in the moist conditions. Tea, coffee and wines all leave dark brown stains on equipment, due to their **tannin** content. A number of special products are sold to deal with this type of discoloration.

However carefully you work, **food spills** are inevitable in the course of cooking. Mild or short term heating turns this material into brown flexible deposits which cling to kitchen surfaces. Any spills on very hot surfaces such as the inside of ovens, rapidly carbonise to black, brittle deposits which bond themselves firmly to the surface. Both of these types of soil need specialised cleaners if they are to be removed effectively.

Types of cleaning agents

The first cleaning agent which comes to mind is water. Water is a good solvent for many substances, but on its own, it is not effective for removing kitchen soil for two reasons:

- It does not wet surfaces well. No cleaning agent can work efficiently if it does not make good contact with the surface which needs cleaning. On hard surfaces such as work benches and crockery, water stands up in rounded globules. This is because it has a *high surface tension*. This tension must be broken if water is to spread out and wet the soiled surface thoroughly. Substances which break surface tension are called detergents or *surfactants* (surface active agents).
- Greasy materials are not soluble in water but the addition of a detergent allows the grease to be *emulsified* into minute droplets which suspend in water and can be washed away.

How detergents work

The molecules of detergents have a head and a tail end. The head is attracted towards water and the tail is repelled by water but attracted to grease. Detergents break surface tension by inserting the water loving (hydrophilic) head end into the surface layer of water molecules. This breaks the strong attraction the water molecules have for one another and so the water is able to spread out and cover the surface.

Fig 12.1 Detergent molecule

Detergent molecules also help to pull grease away from surfaces. The water hating (hydrophobic) tails penetrate the grease whilst the water loving (hydrophilic) heads are attracted to the surrounding water molecules. Agitation of the water in the washing process then helps the detergent to roll up the grease into minute droplets. These droplets are surrounded with detergent molecules and so cannot come together, so they stay suspended in the water as an emulsion.

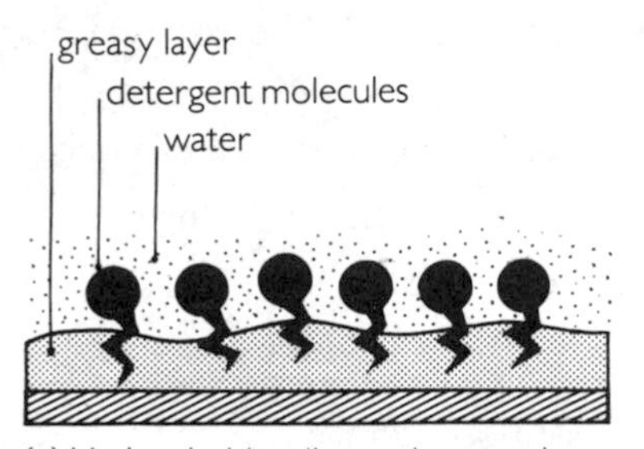

(a) Hydrophobic tails attack greasy layer

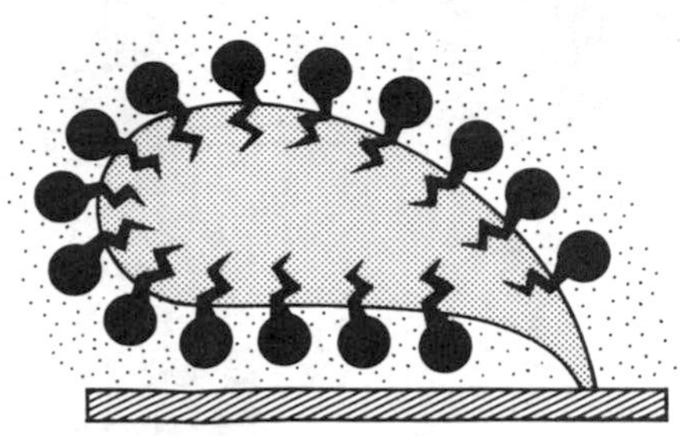

(b) The grease is pulled away from the surface

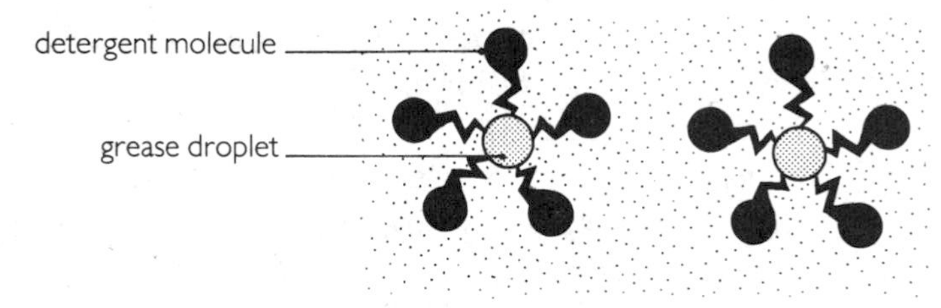

(c) Minute droplets of grease surrounded by detergent molecules. An *emulsion* has been formed

Fig 12.2 How detergents clean greasy surfaces

Exercise: Demonstrating the properties of detergents

Experiment 1

Take two test tubes or small glass bottles.

Fill one two-thirds full of water and, with a dropper, add enough vegetable oil to cover the surface of the water. Place your thumb over the opening of the tube or bottle and shake thoroughly. *Note* the appearance of the mixture. Allow it to stand for several minutes and again *note* the result.

Fill the second container two-thirds full of water and add one or two drops of washing up liquid with a clean dropper. Shake to mix. Add enough oil to cover the surface and shake thoroughly. *Note* the appearance of the mixture and leave to stand.

If you have mixed the oil and detergent in the right proportions, the oil will stay suspended in the liquid and will not separate as in the first tube. The oil has been emulsified.

Experiment 2

Take 2 shallow baking dishes or petri dishes. Grease them lightly. Taking dish 1, use a dropper to place several well separated drops of water on the greasy surface.

Add a few drops of washing up liquid to some water. Drop some of the diluted detergent onto the greasy surface of the second dish.

Leave the two dishes to stand undisturbed for some minutes then answer these questions:

1 How did the drops on the second dish differ from those on dish 1?

2 Can you see any evidence that the detergent is removing the grease?

3 What is this process called?

Types of detergent/surfactants

The oldest type of detergent is soap which is produced by boiling an animal fat or plant oil with a strong alkali such as caustic soda. Soap is still used in personal hygiene but it is not suitable for heavy cleaning jobs in the kitchen. The major drawback of soap is that it does not lather well in hard water and so is a less efficient cleaner when used with water which comes from chalk or limestone districts. Calcium or magnesium salts in these waters react with soap to form a scum which is unsightly and make surfaces slippery.

Modern synthetic detergents are much less affected by hard water salts. There are a number of different types of these surfactants, each with somewhat different properties, which are used in formulating cleaning products to deal with particular types of soil. The various types of detergent/surfactants behave in different ways when they are dissolved in water.

Anionic high foaming surfactants

The active part of the molecule is *negatively* charged or *anionic*. These surfactants are inexpensive and are used in washing-up liquids and general cleaners.

They produce a lot of foam, are mildly alkaline and have good wetting properties. They are used with *hot* water for general cleaning jobs where there is no heavy soil.

Cationic detergent/ sanitizers

The active part of the molecule is *positively* charged or *cationic*. The most commonly used cationic detergents are Quaterary Ammonium Compounds, often known as QACs or QUATs.

They are high foaming surfactants with only mild detergent power, but they have the ability to kill bacteria so are used in kitchen sanitizers.

Note: Cationic (+) detergents must not be mixed with anionics (-) because they will inactivate one another.

Non-ionic low foaming detergents

Non-ionic detergents do not ionise when in solution so have *no charge*. The whole molecule is active in cleaning. Non-ionics, however, are more expensive to produce than anionics so are used only when their special qualities are required.

They are neutral in pH, low foaming but emulsify greases well. They are used in floor cleaning machinery where a high foaming product would be undesirable. As they have no charge of their own , they can be mixed with either anionics or cationic detergents.

Amphoteric detergents Amphoteric detergents are affected by the pH of their surroundings so they have the properties of cationic detergents in acid conditions, non-ionics at neutrality and anionic detergents when the solution is alkaline.

The molecules of these detergents carry positive or negative charges depending on the pH of their surroundings.

Acid	*Neutral*	*Alkaline*
+ charge	No charge	– charge
Cationic	Non-ionic	anionic

Amphoterics are used in some oven cleaners.

Characteristics of cleaning agents

A cleaning agent has to be able to perform a number of different functions. It must be able to:

- wet the surface which needs cleaning.
- penetrate any problem soil.
- lift the soil from the surface.
- suspend the soil and prevent it redepositing on the cleaned material.

A cleaning product has also to be tailored to:

- the type of soil to be removed e.g., grease, scale, rust.
- the nature of the equipment/surface to be cleaned e.g., silver, steel aluminium, plastic.
- ensure the safety of the people who will use the product.

To perform all these tasks, the detergents used for kitchen cleaning consist of:

- One or more surfactants to remove grease and wet out the surface.
- A number of additional substances or 'builders'.

Types of builders

Alkalis Alkalis are used in products designed to remove greasy food residues. They vary from mildly alkaline substances such as sodium bicarbonate, to strongly alkaline chemicals such as washing soda (pH 13) and caustic soda (pH 14). Alkalis are good cleaning agents in their own right and help to turn grease into soap which rinses away easily.

Sequesterants These are substances which keep hard water salts out of circulation so that they do not interfere with detergent action or deposit as scale in dish washers or other machines. The most common sequesterants are phosphates which have the additional advantage of keeping the soil suspended so that it does not redeposit on the cleaned surface.

Solvents Products designed to remove greasy deposits contain non-watery solvents in addition to surfactants, to assist the removal of these problem soils.

Acids Acids may be used on their own or in addition to surfactants to remove scale from bain maries, steamers and kettles. Organic acids such as citric and oxalic acid are used with surfactants for regular descaling of catering equipment, while hydrochloric acid is employed in heavy duty descaling fluids.

Bleaching agents Bleaches are used to remove stains such as tannin – which causes the brown tea and coffee stains. Bleach also whitens fabrics which yellow in sunlight. The most common bleaches are laundry borax (sodium perborate) and household bleach (sodium hypochlorite).

Sodium perborate is a mild bleach which works only at high temperatures. It is present in laundry powders and some preparations sold for removing tannin stains.

Sodium hypochlorite is a very effective bleach which is active at low temperatures. It not only bleaches but is also a powerful bactericide. It is the basis of many toilet cleaners and is incorporated, in solid form, in some scouring powders, machine dishwashing powders and sanitizers. It has, however, some disadvantages:

- It is dangerous to handle in a concentrated form (*see* p 169)
- It does not remain active for long once it has been diluted with water.

For these reasons, modern cleaning products tend to employ hypochlorite bleaches less than in the past and to rely on QACs for sanitizing surfaces.

Selecting cleaning agents

Specialist suppliers make a wide range of cleaning products for use in food premises. Managers have to decide which of these products are needed for their particular situations. It is as well to limit the number of cleaning agents used and so ease the burden of training employees to use them safely and effectively. Details of common products, likely to be needed in most types of food premises, are given below.

Neutral general detergents

These are blends of anionic and non-ionic surfactants suitable for handwashing of crockery and glassware. They can also be used for washing floors and walls.

All purpose cleaner/detergent degreasers

These are mildly alkaline preparations containing anionic and non-ionic surfactants, solvents and sequestering agents. They can be used on walls and floors. They are safe to use on tiles, painted surfaces and steel equipment but may dull aluminium slightly.

Alkaline detergent for dishwashers

These are highly alkaline preparations with hypochlorite bleach for stain removal and disinfection. They contain a sequesterant to enhance detergent action and prevent scaling. They are not suitable for washing equipment composed of aluminium, zinc, or alloys of these metals.

Rinse aids

These products are formulated to complete the cycle in dishwashers so that the glass or crockery emerges sparkling and smear free. They contain a low foaming surfactant and an alcohol to aid evaporation.

Sanitizers

These are generally based on a blend of a non-ionic surfactant and a QAC. They can be used on all types of kitchen surfaces, tableware and utensils. Heavily soiled articles should be cleaned with a neutral detergent and rinsed before the sanitizer is applied. For adequate bactericidal action the sanitizer should be in contact with the surface or equipment for at least 10 minutes as QACs are not rapid acting agents.

Descalers

These consist of an acid, often phosphoric and a non-ionic detergent. They remove hard water scale from bain maries and steamers, and milkstone from drink dispensers. Suitable for steel, glass and copper surfaces but not for plastic kettles.

Oven decarbonisers

These are highly alkaline products containing detergents and sequesterants. They are used to remove baked-on carbonised food deposits from ovens and grills. They can be used with boiling water to remove the heavy deposits of grease from fat fryers.

Abrasive cleaners

Abrasive powders and creams are used to remove stains from porcelain and enamel surfaces. The products contain substances with varying degrees of abrasiveness according to the surfaces they are designed to clean, e.g. pumice, felspar, and chalk. In addition there is usually a bleach to oxidise stains, a sequesterant to soften water, and a detergent to cut through the grease.

Exercise

Match the type of soil with the best kind of detergent.

1 Grease on pans
2 Limescale build-up in steamer
3 Dirty porcelain sink
4 Baked-on grease in oven
5 Clean equipment needing sanitizing

(a) Acid descaler
(b) Abrasive cream
(c) Decarboniser
(d) Neutral detergent
(e) Detergent/QAC
(f) Rinse aid

Using cleaning agents

When you have chosen the appropriate detergent for a cleaning operation, the next point to consider is how to use it to remove the soil. Some cleaning agents are best used straight from the bottle while others need to be diluted. Washing up liquids and general cleaning detergents are usually purchased in concentrated form and have to be diluted in water for use. Too little detergent in the water will make them ineffective; too much will be uneconomical, may dry the user's hands or, in time, even damage some surfaces. So it is important to *measure* these substances *accurately*.

Some manufacturers make dilution easy by supplying their products with dosing caps which dispense the quantity of detergent appropriate for a particular volume of water e.g. one measure to five litres of water. Alternatively, containers of detergents can be connected to tap proportioners which deliver the appropriate amount of cleaning agent for the volume of water, automatically. Large catering establishments often have cleaning stations where the commonly used detergents are delivered at the correct strength.

Products such as window cleaners and oven decarbonisers are best applied neat. In some cases this can be done simply with a cloth but in many cases a jet or spray is more effective. Manufacturers often supply these products with directional jet caps or with leads to a spray gun.

Safe use of cleaning agents

Many cleaning agents are potentially dangerous because they are caustic or corrosive to living tissue or toxic if swallowed or inhaled. It is therefore *essential* that employees know the hazards of any products they may use in the course of their work.

Both manufacturers and employers have *legal obligations* to promote safe handling methods and to minimise the risk of using hazardous chemicals. Manufacturers try to formulate their products so they cause as little risk to the user as possible. However, stubborn soils often necessitate the use of dangerous chemicals and in these cases manufacturers are obliged to include hazard warnings on all containers of these products.

Two types of information have to be given:

1 Details of the risk involved, for example:
- irritating to the skin
- damaging if splashed into the eyes
- toxic if inhaled

2 Advice on how to minimise the risk, for example:
- use rubber gloves
- use face mask or goggles
- only use in good ventilation

Statutory requirements

Under the Control of Substances Hazardous to Health Regulations 1988 (COSHH) employers must:

1 *Assess the risk to health of their workers of any substance they handle in the course of their work*. This means they (or someone they appoint) must familiarise themselves with the products handled by their employees and find out whether any of them present a hazard.

2 *Introduce measures to prevent or control the risk of handling hazardous chemicals*. This may be done by changing a hazardous preparation for one with minimal risk e.g. replacing a sanitizer based on household bleach to one containing a QAC or, if this is not possible, they must provide the protective clothing or equipment to reduce the danger.

3 *Inform employees of the risk of hazardous substances*. This may be in the form of a memo which the employee is obliged to read and sign, or a notice in the part of the workplace where the product is used.

4 *Train employees to work safely*. New workers must be shown how to use the products and not left to work unsupervised until they can use them safely.

5 *Ensure that hazardous chemicals are used and stored correctly*. Hazardous chemicals should be stored in *locked* cupboards or store rooms. If any agents are bought in bulk and distributed in smaller quantities, the containers issued to staff must be labelled with the nature of the product and any warning notices which were on the original containers.

Which cleaning agents are hazardous?

Low hazard cleaning agents

Neutral and mildly alkaline detergents such as washing up liquids, general purpose cleaners and non-bleach sanitizers come into this category. The only adverse effect of these products is to dry the skin if used for prolonged periods. You should always rinse and dry you hands after using these detergents. If, in spite of this precaution, you find your skin becoming dry and cracked, you should consider using rubber gloves or applying a suitable barrier cream before commencing work.

Acidic preparations

Preparations sold for brightening silver and brasswork, for descaling equipment, and some toilet cleaners contain acids.

All acids are corrosive to the skin and can cause serious damage if splashed into the eyes. The degree of hazard will vary with the type of acid present in the product so the manufacturer's instructions should be followed *precisely*.

For example, heavy duty descaling liquid contains hydrochloric acid (strong inorganic acid) and precautions must be taken:

- Ventilate the area well
- Wear rubber gloves, overalls and face protection
- Do not inhale the acid vapours

Silver brightening liquid is mildly acidic and precautions for use are:

- Rinse and dry hands after use.
- Use gloves for prolonged use.

Alkaline products

Alkalis are caustic and damage the skin and eyes. Again the degree of hazard varies according to the type of alkali employed. Detergent degreasers are usually mildly alkaline, while products sold for cleaning ovens and fryers are highly caustic as you can see from the information on the labels below.

Oven/fryer degreaser – Highly caustic
1 Ventilate the area well
2 Wear heavy duty gloves and face protection.

Detergent degreaser – Non-caustic alkali
Use gloves for prolonged contact.

First aid for acids and alkalis

If splashed on the skin:
Wash the affected area *immediately* with plenty of water. Remove any contaminated clothing and wash any skin which has been in contact with the acid or alkali. Continue washing the splashed area for at least 10 minutes. Seek medical advice if irritation persists after treatment.

If splashed into the eyes: Wash the eyes gently but thoroughly with running water from a tap or eye-wash bottle. Alternatively, fill a sink or large bowl with water, immerse face and open and shut eyes to assist removal of the substance.

Seek medical advice as to whether further treatment is necessary.

If swallowed: Give plenty of water to drink. **Do not** make the person vomit. Seek medical attention.

Solvent based cleaning fluids

Some cleaning fluids contain solvents which catch fire at quite low temperatures. These substances must be stored in cool conditions away from direct sunlight. They must **not** be used near naked flames.

These products should only be used in well ventilated areas since many have a narcotic effect if breathed in high concentration. When spraying these products or if splashing is likely, use rubber gloves and eye protection.

First aid

- Wash away splashes from the skin with soap and water. Remove contaminated clothing and treat skin beneath.
- If eyes are affected, treat as for acids and alkalis.
- If swallowed give plenty of milk. Do not induce vomiting.
- If the person feels dizzy or sick, take them into fresh air. Keep them warm and let them rest.

Bleaches

Preparations containing household bleach, e.g. Domestos, Parazone and similar products, must be handled with care. The bottles must always be stored upright and tightly stoppered. Care must be taken to prevent the liquid splashing into the eyes. Bleaches must **not** be mixed with other cleaning products (*see* below). First aid as for acids and alkalis.

Mixing of cleaning agents

The simple rule is, **do not mix chemical cleaners together**. In some cases, mixing produces dangerous substances:

- Hypochlorite bleach mixed with acid preparations produces chlorine, a choking gas which can seriously damage the lungs. Remember that some toilet cleaning products contain acids so *never* use more than one type of toilet cleaner at a time.
- Some liquid cleaners contain ammonium compounds which help to remove grease. The ammonia vapour has a pungent odour and convinces the user that the product is powerful. If these substances are used at the same time as a hypochlorite bleach, a chemical reaction occurs producing chloramines. These acid fumes cause choking. Luckily the effect, though unpleasant and best avoided, is only temporary and does not cause permanent ill effects.
- Any mixing of acid products with alkalis, creates *heat* and in some cases the reaction could be violent enough to spray the corrosive liquid in all directions.

The simple and obvious answer is not to mix different types of cleaning agents.

Task

Collect together as many different cleaning agents as you can. Study their labels and any advertising literature which is available. Fill in the information you obtain on a chart similar to Table 13.1

Table 13.1 Uses and properties of cleaning agents

Product name and type	*Uses*	*Neutral/ acid alkaline*	*Solvent flammable narcotic*	*Method of application*	*Hazards*	*Precautions*
Tig 3J Washing up liquid	Hand dishwashing Washing hard surfaces	Neutral	None	Dilute to use for washing dishes	Skin drying on prolonged use	Use barrier cream or gloves
Domestos bleach	Toilet cleaning. Disinfection	Alkaline	None	Use direct from bottle	Damage to skin and eyes	Use goggles and gloves

Exercise

1 Name a pair of products in your table which would produce heat if mixed together by an untrained operative.

2 Name a pair of products which would produce chlorine if used at the same time.

3 Name a product containing a flammable solvent. How should it be stored? What precautions should be taken in use?

4 Describe the first aid required if a colleague splashed oven cleaner into his or her eyes.

Hygiene in food premises

Good hygiene involves two processes which are different but interconnected:

1 *Cleaning*. As we have seen earlier in the chapter this achieved by using the correct detergents in well designed cleaning schedules, with regular audits to check that high standards are maintained.

2 *Control of bacterial growth*. In Chapter 3 we saw how important it was to control the spoilage organisms which deteriorate food and the pathogens which cause food illnesses. The main method of controlling bacteria is through **time** and **temperature** but disinfection and sanitizing also play a part.

There are a number of terms which are used in connection with the elimination or control of micro-organisms. They are sometimes misunderstood or used incorrectly. The following definitions should clear up these difficulties:

- **Sterilisation** means to kill or remove all micro-organisms.
- **Sterilant** This term is applied to powerful disinfectants which will sterilise if used according to the manufacturer's instructions.
- **Disinfection** means to kill microbes but not necessarily bacterial spores.
- **Antiseptics** are disinfectants which are safe to use on human tissue.
- **Bacteriocides** are substances capable of killing bacteria.
- **Bacteriostats** prevent bacteria from multiplying but do not kill them.
- **To sanitize** means to reduce the load of micro-organisms to an acceptable level.
- **Sanitizers** are usually combinations of detergents and disinfectants which allow surfaces/equipment to be cleaned and sanitized in one operation.

True sterilisation, as in a hospital operating theatre, is rarely achieved or necessary in food handling. The only situations where food is likely to be sterilised is in very high temperature processes such as deep frying or pressure cooking. To keep food in hygienic conditions it is only necessary to *reduce* the micro-organisms to such small numbers that there is little danger of them multiplying and causing spoilage or food illness.

Disinfectants

Few types of chemical disinfectants are suitable for use on food contact surfaces because most of them taint food. They may:

- leave a strong smell or taste on the surface
- leave toxic residues behind
- discolour surfaces.

Because of these drawbacks, many excellent disinfectants are only suitable for use *outside* the kitchen to disinfect toilets, drains, waste bins, etc., or in diluted form as antiseptics for personal use.

Disinfectants used in the food industry

Sodium hypochlorite solutions

Examples of these products include 'Bleach', Milton, Parazone, etc. Until quite recently these preparations were used extensively in the disinfection of food premises.

Advantages These products are inexpensive. They remove stains and are very effective in killing bacteria and viruses. They have a pronounced smell but it disappears quite quickly.

Disadvantages They do not wet surfaces well and need careful handling in concentrated form. Their activity lasts only for a short time.

Iodophors

Examples include Betadine and Pevadine.
These are complexes of detergents and iodine.

Advantages They are active against a wide range of micro-organisms and this action is maintained for long periods.

Disadvantages They attack some metals and are expensive to use.

QACs (QUATs)

These include Task and Savlon and are also incorporated in sanitizers. As mentioned earlier, these are cationic detergents.

Advantages They are non-toxic, odourless and stable to heat. They are suitable for use on both steel and plastic surfaces.

Disadvantages They are not quick acting and do not kill bacterial spores. They are inactivated by the presence of organic matter (e.g. food). They are expensive compared to hypochlorites.

Sanitizing in food premises

Heat, particularly *moist* heat is the best sanitizer in the kitchen. Plenty of *hot* water frequently and vigorously applied will usually reduce the load of micro-organisms to a safe level and leave no dangerous chemical deposits behind.

In those areas where there is a high risk of contamination such as mincers, blenders, food slicing machines, chopping boards, hot cupboards and display units, the use of a sanitizer is appropriate. In the case of machines, the equipment must be turned off and dismantled. Any food residues must be removed and then the sanitizer applied. To ensure killing the maximum number of bacteria, immerse the parts in the sanitizing solution for at least 10 minutes. Afterwards rinse well and allow to air before reassembling the equipment.

Storage of cleaning products

Cleaning materials and equipment must not be stored in any food room. They must be kept in a separate store used only for this purpose. It should be fitted with slatted shelves and be well ventilated, cool and dry. Any cleaning agents and chemicals which require specialist training for safe handling, must be kept in a separate, locked area and issued only to those responsible for their use.

Gas bottles

Gas bottles used in the catering industry range from the small cylinders for gueridon lamps to the very large cylinders of Calor gas employed for outside catering functions.

The small cylinders may contain butane or liquid petroleum gas. These may be stored in the room with the cleaning products. The Calor gas cylinders must be stored *away* from the building in a specially constructed compound which is kept locked.

Cleaning staff

Cleaning is a very important part of the work of any efficient catering establishment. Staff induction schemes should make this abundantly clear so that personnel realise how essential it is and take a pride in their work. Some firms run incentive schemes to encourage this attitude and award bonuses for the best kept areas on the basis of periodic inspections.

Cleaning schedules are one means of ensuring that cleaning is carried out efficiently and at the correct intervals.

Drawing up cleaning schedules

There are no hard and fast rules for drawing up schedules as every establishment is different. The main objective is to ensure that the area is *thoroughly* cleaned over a set period e.g. two weeks. Here is some general guidance as how to approach the task:

- Itemise all the tasks involved in cleaning the areas over a set period.
- Break down the areas involved into groups according to the amount of *use* they get and their *importance*. For example, tables and working surfaces are used constantly and have direct contact with food so they must be cleaned frequently throughout the day – shelves are not in direct contact with food, so weekly cleaning will be sufficient.
- Consider the staff available – their experience and abilities.

When the schedule is complete, it must be displayed in a prominent place so *all* staff are aware of their responsibilities. Some establishments break the schedules down and allocate them in the form of personal job cards to each member of the cleaning staff.

The following schedule for a small canteen servery and production area shows how tasks are broken down and allocated to staff.

CLEANING SCHEDULE

Daily	**Staff responsible**
Servery	
• Wash and sanitize all table tops and work surfaces	X
• Wash and sanitize all table legs and undershelves	X
• Wash, dry and polish all sneeze guards	X
• Wash and sanitize service counter and shelves	X
• Thoroughly clean hot cupboards, polish doors	X
• Wash heat lamp bulbs, dry and polish	X
• Wash and dry cold food well	X
• Wash and dry tray service rack and tray service counter	X
• Wash, dry and polish panelling of food service counter	X
• Wash, dry and polish drinking water area	X
• Wash and clean out drip tray of the still and clean all steam nozzles and sterilise	X
• Clean griddle plate, oil slightly	X
• Strain deep fryer, remove all debris, wipe and refill with oil	Y
• Clean boiling ring, remove all spillage	Y
• Clean microwave oven turntable and grid – polish	X
• Clean and polish coffee machine and clean coffee jugs thoroughly	X
• Remove all rubbish bags from steel	

CLEANING SCHEDULE continued

cupboards, wash and sterilise inside and out. Polish outside	Y
• Scrub floor removing all grease and food from around table legs and feet	Y
• Scrub under service counter	Y
• Sweep floor in food store, wash if necessary	X
Kitchen	
• Wash and sanitize all table tops and white chopping boards	X
• Wash and sanitize all table legs, sink legs and undershelves of stoves and tables	X
• Clean dish washer, remove grills and wash thoroughly	Y
• Wash potato machine, empty and wash peel trap thoroughly	Y
• Clean the boiling table. If necessary remove surrounds and clean. Clean drip tray.	Y
• Scrub floor removing all grease and food scraps around feet of tables	Y
• Clean all sinks, strainers and stoppers	Y
Twice weekly	
• Spray ovens with oven cleaner and clean thoroughly. Clean all racks and drip tray. Wash, rinse and dry. Note: *if necessary clean more frequently*	Y
• Empty and clean equipment. Reline drawers and replace equipment	X
• Dismantle boiling table and clean all pieces thoroughly. Replace ready for use	Y
• Remove deep fryer and clean surround, replace unit. Remove boiling rings and thoroughly clean surround. Replace unit	Y
• Remove griddle plate and thoroughly clean surround. Replace unit.	Y
Weekly	
• Defrost and wash out all refrigerators	X
• Drain deep fryer, fill with suitable cleaner and boil to remove grease. Wash, rinse, dry and refill	Y
• Wash all shelves in food store	X
Fortnightly	
• Remove all ventilation and extractor grills, wash thoroughly and replace	Y
• Wash all walls in the kitchen and serving area	Y
• Wash all ventilation hoods over cooking equipment	Y

Note: The deep cleaning of ventilation hoods, canopies and extractor grills and fans may be carried out by specialist cleaning companies. Up-to-date kitchens have a computerised cleaning system which cleans the ventilation ducts automatically at a pre-set time.

Task

Draw up a cleaning schedule for the establishment where you work so that it will be cleaned thoroughly over a period of one month.

13 Pests and pest control

Wherever there is food, warmth, and shelter other creatures besides ourselves will make themselves at home. Kitchens and food stores make ideal refuges where rodents, birds and insects can find nourishment and, in some cases, take up residence and raise their young. These visitors are unwelcome in food premises for a number of reasons:

- They cause economic loss by spoiling a great deal of food – far more than they actually consume.
- They carry pathogenic organisms in or on their bodies so transmit disease when they contaminate food.
- Some pests, such as rodents, cause damage to buildings. The initial damage may be small, but a few feet of stripped electric cable may start a major fire, or a chewed water pipe, an extensive flood.

Managers of food businesses have a legal responsibility to 'prevent, so far as is reasonably possible, entry of birds and any risk of infestation by rats, mice or insects', under the Food Hygiene (Ammended) Regulations 1989 (*see* p 20). To fulfill this obligation, they have to ensure that their buildings are as pest proof as possible and are maintained in good order. They also need to maintain a liaison with the pest control officers of the Local Authority and possibly a commercial pest control organisation, so that they have expert advice and assistance in preventing and, when necessary, controlling infestations in their premises.

Employees also have a vital role to play in safeguarding food stocks and preventing infestations in their kitchens. Food handlers are the eyes and ears of the management; their vigilance in observing and reporting the presence of pests can prevent a small problem from turning into a major infestation requiring expensive irradication measures and even the closure of the premises by the EHOs. Training is essential. Staff need to know:

- Where food pests are likely to be found
- The signs which indicate the presence of pests
- The methods of goods housekeeping and hygiene which make kitchens and store rooms less attractive to pests.

Rodents

Rats and mice belong to a group of mammals whose teeth and jaws are specially adapted to gnawing. They need to gnaw hard materials constantly in order to wear down their chisel-like incisor teeth which grow throughout life. As a consequence, they do a great deal of damage to buildings by gnawing through wooden furniture and shelving, stripping cables, and damaging water pipes. They damage a great deal more food than they consume; wandering from one packet to another nibbling a hole here and there. Apart from this physical damage, they contaminate food with their smell, hair, droppings and urine.

Rats can transmit a number of diseases. The most important is *Salmonella* poisoning but they have also been implicated in the transmission of Weil's disease (a severe fever and jaundice), trichinosis, rabies and other diseases.

There are two species of rats which infest food stores in the UK, the Brown rat, *Rattus norvegicus* and the Black rat, *Rattus rattus*. The Brown rat is the larger of the two, with a body length of about 250mm and weighing up to 500g while the Black rat is around 200mm long and weighs about 225g. Both types of rats produce 3–5 litters of young a year, each litter numbering 6–8 young.

Rats are intelligent and agile creatures. They will climb, burrow and even swim to gain access to a building where there is food and shelter. The first defences against rats are secure and well built premises. Rats can enter through surprisingly small crevices, so all holes in walls around pipes and cables must be sealed, and drains and floor gulleys must have effective water seals. Drainpipes should be fitted with wire balloons inside and cone guards on the outside so that rats cannot use these structures as a means of access to the building. Doors and windows should fit tightly and any external wooden doors should be fitted with metal kick plates so that the animals cannot gnaw their way in. Air bricks and ventilation grids should be protected with metal mesh if the holes are bigger than 6mm across, so as to exclude mice as well as rats.

The surroundings of food premises should be inspected regularly to make sure that no rodents are in outbuildings and that no food debris has accumulated which will attract the animals to the premises.

Within the building, food should be stored as far as possible in bins or other rat proof containers. When food is supplied in sacks, these should be inspected when they are first taken in and moved at intervals so that they do not provide nesting sites for rodents.

In any building there are areas which are rarely visited, for example service lift shafts, lofts and cellars. A member of staff should be detailed to inspect these places regularly to look for signs of infestation.

Rats are nocturnal creatures so staff are unlikely to detect the animals themselves, only the signs of their presence:

- nibbled packets or sacks of food
- foot prints in spilled flour or other fine powders
- the characteristic odour left on contaminated food
- hair adhering to food or fittings
- 'loop' marks where the rodents' greasy coats have brushed against surfaces

- droppings on the floor where rodents have been feeding
- gnaw marks on woodwork or food containers

Any of these signs indicate that rodents are resident or visiting food premises and indicate that a search for their mode of entry is *urgently* needed.

House mouse

The House mouse, *Mus musculus*, is much smaller than the rat; up to 80mm long and 13–20g in weight. Nevertheless it can cause considerable damage to buildings and foodstores. Mice are very prolific breeders, having 10 litters a year of up to 16 young, so a small infestation can become serious within a short period. They are more difficult to exclude from buildings than rats as they can squeeze through any hole larger than 6mm across. They are also more difficult to catch since they feed sporadically rather than regularly as rats do.

Irradiacation of rodents

The two main methods of killing rats are the use of traps and poisoned baits. Rats are creatures of regular habits; they tend to use the same runs to reach food night after night. Baits or traps laid in these runs stand the best chance of catching the rodents. However, rats are suspicious of any unusual objects and they may avoid baits for several days before their natural curiosity overcomes their fear of the unknown. In catering premises, the bait must be at least as attractive as regards taste and smell as any food in the surrounding store, otherwise it may be ignored.

Warfarin was the first widely successful rodenticide which was not an acute poison and so safe to use in catering premises. It acts as an anticoagulant preventing the blood from clotting so causing haemorrhages, and eventually the death of the animal. The warfarin baits need to be replaced on several consecutive nights until the rats have consumed enough of the poison to kill them.

In controlling pests, battles are won but there is never a complete victory. After some years of success with Warfarin, reports of 'super rats' began to appear in some parts of the UK – rats which were Warfarin resistant. Since then, new anticoagulant rodenticides such as difenacoum and bromadiolone have been developed which are effective against Warafin resistant rats and mice. Some like Storm (ICI), containing Flocoumarin, can kill rodents in a single feed and so are more convenient and economical than the original type of bait. Another rather surprising substance in the armoury of the rat catcher is Calciferol. This is more familiarly known as Vitamin D. It is used in much larger amounts than usually present in the animal's diet and causes excessive amounts of calcium compounds to be laid down in various parts of its body.

Mice The anticoagulant preparations and Calciferol are used to deal with mice as well as rats. In addition, Alphachoralose is sometimes employed. This lowers the body temperature so that the mice lose consciousness and die.

Birds

Sparrows and wild pigeons are the types of birds which cause problems in catering premises. Like rodents, they are attracted to food stores and waste disposal areas where there is food available and suitable sites for building nests.

Birds feeding in rubbish areas leave behind droppings, feathers and twigs – all very messy and insanitary materials. Birds carry *Salmonellae, Campylobacter* and other pathogenic bacteria in their intestines, so can infect any food to which they have access. Their feathers are infested with insects and mites, which if they get into food stores can be difficult to irradicate.

The droppings of birds, particularly pigeons, cause considerable damage to the external fabric of buildings. Fungi grow on the accumulations of bird excrement, creating acids which eat into the stone or brickwork.

As with most infestations, good housekeeping is the best preventative. If waste areas are kept clean and bins tightly covered, there is less likelihood of sparrow invasions. Ventilation grids and louvres which might provide easy access for birds should be covered with wire grids.

As regards pigeons, the main defence is to deny them space to perch upon. This can be done by covering ledges with thick gels or by installing spring tensioned horizontal wires so that pigeons cannot balance on the ledges and therefore choose other roosting sites.

Insects

Houseflies and blowflies

The common housefly *Musca domestica* and its larger cousin the blowfly *Calliophora vomitoria* are the most dangerous insect pests that frequent catering premises. Their rapid breeding rate and insanitary feeding habits make them major agents of transmission of pathogenic organisms to food.

Life cycle of the fly Flies lay their eggs in moist decaying matter such as refuse heaps during the summer months. Five or six batches are laid in a season with around 120–150 eggs in each group.

The eggs hatch in about 12 hours and the resulting maggot seeks a moist dark place to feed voraciously, moult twice and then pupate.

Inside the pupal case, the organs of the maggot break down and rearrange themselves into those of the adult fly. The length of the life cycle varies with temperature, in the cool early summer it may be as long as 45 days but in the hottest part of the year it may be as short as 10 days.

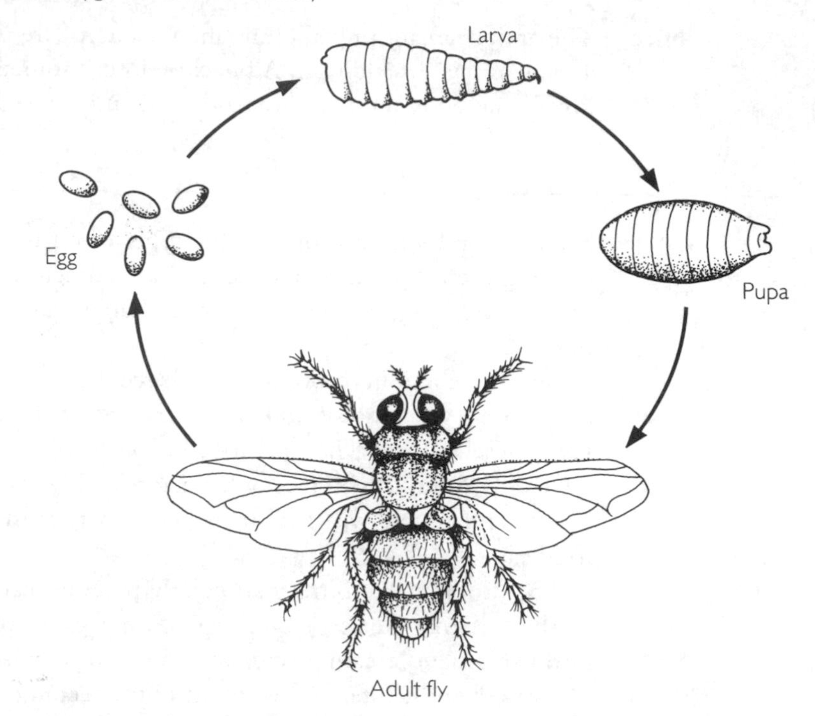

Fig 13.1 Life cycle of a house fly

The adult fly

The feeding habits of flies seem designed by nature to contaminate food with pathogenic organisms.

- They are not particular where they feed. They may feed in a dust bin or manure heap and then transfer their attention to food in a kitchen if it is accessible.
- Flies have very hairy bodies and legs and are equipped with sticky suckers on their feet. This means they carry a large burden of bacteria on the outside of their bodies as well as in their intestines.
- Flies can only take in liquid food. When feeding, they place their sucking tube (proboscis) against the food and release saliva to digest the meal. A portion of the previous meal is often voided with the saliva. This unsavoury method of feeding means that the fly which lands on your meat, may well deposit on it, the remains of a previous feast on a nearby manure heap.
- Flies defaecate while feeding and their faeces carry *Salmonellae,* other pathogenic organisms, and the eggs of parasitic worms.
- Food can be contaminated by fly debris such as pupal cases and dead flies – which are just as dangerous as the live insects.

Prevention and control of flies

- The first defence against flies is to remove the places where they are likely to lay their eggs. Rubbish and manure heaps should not be near buildings where food is prepared or stored. Dust bins should be tightly covered and the refuse area kept clean to avoid attracting the insects.

- The hot summer months, when flies breed, are the time of year when windows are likely to be thrown open for ventilation, but also allowing flies to enter. Fly screens are the answer to this problem. They can be hinged into existing windows or fitted into sliding runners so they can be removed when no longer needed in the cooler months.

Fig 13.2 Fly screen *Courtesy: Rentokil*

- Within premises there are two main methods of killing flies:

1 Many shops, supermarkets and kitchens use insectocutors to kill flying insects, including flies. The devices attract the insects to an ultra violet lamp and then stun them electrically. The equipment should be

positioned so that neither staff nor food will be irradiated by the lamp. The trays should be emptied regularly so dead insects do not accumulate.

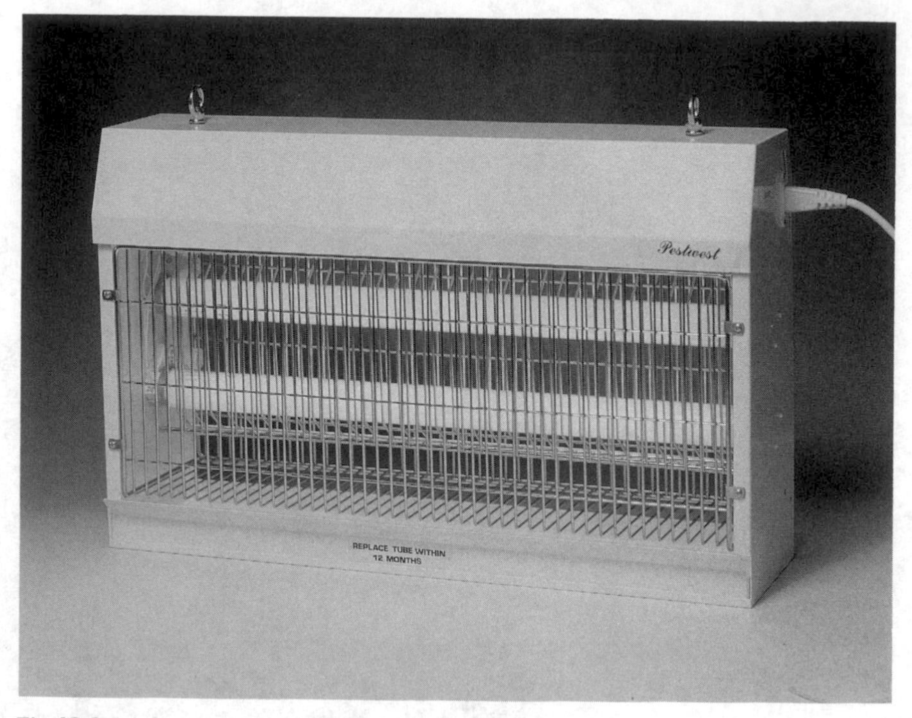

Fig 13.3 Insectocutor *Courtesy: Killgerm Chemicals Ltd*

2 Chemical methods of killing flies have to be approached with caution because of the danger of contaminating food. The safest and most effective products are synthetic pyrethroids such as Alphacypermethrin. This insecticide can be supplied as a wettable powder which suspends, evenly in water and leaves an even layer on the surface after the water has evaporated. This is applied to such surfaces as wood, brick and glass frequented by flies and other insects. It is non-staining, odourless and of low toxicity to man but lethal to insects within a matter of minutes. It retains its activity for prolonged periods so there is no need for frequent applications.

Cockroaches

Cockroaches belong to a group of insects with flattened bodies and long thread like antennae. They are nocturnal creatures, hiding away during the day behind skirting boards, under cookers, in ventilation ducts and other near-inaccessible places. They will eat almost anything – paper, clothing and leather – but are especially attracted to sweet and starchy materials.

Two types of cockroach infest buildings in the UK, the dark brown Oriental cockroach *Blatta orientalis* and the lighter coloured German cockroach, *Blattella germanica*.

Cockroaches lay their eggs in purse shaped cases. *B. orientalis* deposits hers in crevices as soon as they are formed, but *B. germanica* carries the cases around on her body until the eggs are ready to hatch. The eggs change into worm shaped nymphs which feed and moult and turn into miniature versions of the adult insect.

Cockroaches spoil a great deal of food, leaving behind them a characteristic 'tom cat' smell. They thrive on food debris in the dusty areas of buildings, so carry an array of pathogenic bacteria, including those which cause food poisoning. Significant numbers of cockroaches in food premises are viewed by the EHOs as indicating a poor level of maintenance and hygiene, and are likely to lead to further investigation. In severe cases, cockroach infestation may even lead to closure of a food business or hospital kitchen.

Control of cockroaches

As with many other pests, prevention is much better than cure. If a building is well maintained there will be fewer crevices to afford hiding places for the insects. Likewise if stoves and other machinery are installed in such a way that all parts are accessible for cleaning and floors are washed regularly, there will be little food available to tempt the insects to take up residence.

Insecticidal lacquers and sprays

Insecticidal lacquers can be used to produce a barrier between the areas where the insects might hide and their possible feeding sites. Synthetic pyrethroids such as Fendona (Shell Research), containing Alphacypermethrin, are active against cockroaches as well as flies and can be sprayed or painted onto skirtings, door and window frames, and entry points such as water pipes or gullies. The insecticide crystallises on the surface and stays active for prolonged periods.

Roach traps

Because cockroaches hide away during the daylight hours, it can be difficult to gauge the extent of any infestation problem which may be present in a building. The answer to this problem is to use roach traps in likely places in the premises.

These traps are made of cardboard, either in a tent shape or, for confined spaces, in a low line version. Inside the trap there is a strip of non drying adhesive, a tablet containing a food bait, and a chemical substance which attracts the insects. The attractant is a pheromone, one of a biologically active group of chemicals which insects use to communicate with each other. Chemists have been able to synthesise the aggregating pheromone which causes cockroaches to crowd together with others of their species. The insects are therefore trapped by their own body odours!

The traps can be laid in suitable areas and examined at intervals to gauge the extent, if any, of cockroach infestation. If only a few cockroaches are found, the traps alone will be sufficient to keep the problem under control. If the numbers trapped, indicate a more serious infestation, the area can be sprayed and further traps laid to monitor the success of the operation.

Fruit flies and vinegar flies

There are a number of related species of these small flies which are drawn to rotting fruits, fermenting liquids and unwashed milk containers. The fruit fly, *Drosophila melanogaster* is the best known of these because it has been used for many years in laboratories for genetic experimentation.

Fruit flies are characterised by their bright red eyes and their slow hovering flight. They breed extremely quickly in hot conditions, so can soon reach plague proportions when they get into kitchens and food stores. The adults are known to obtain much of their moisture from faecal material so, although they have not be shown to spread intestinal diseases, it is highly likely that they are capable of doing so.

Control The containers where the flies have been breeding should be washed thoroughly and any over-ripe fruit or other suitable breeding materials, cleared out. A pyrethroid spray can be used providing foods are removed or protected from contamination.

Beetles

These insects are easily recognised by their horny wing cases which cover the delicate flying wings when they are not in use. A number of species of beetles infest stored food products.

The Biscuit Beetle

The Biscuit Beetle is a reddish brown beetle only 2–3.5 mm in length. Despite its name, it thrives on flour, bread, soups powders and spices as well as many non-food items such as paper – and even poisonous chemicals such as strychnine.

The female lays about 100 eggs in suitable foodstuffs or in crevices nearby. The eggs hatch into larvae which explore their surroundings actively. As they are very small, they can easily squeeze their way into packets of food. The larvae feed and moult four times, and then pupate in a cocoon made from food particles glued together with their own saliva. After 12–15 days, the adults chew their way out of their resting places.

Control Any food found to be infested must be cleared out and the store cleaned *thoroughly*. A puffer pack of insecticide containing Bendiocarb can then be applied to cracks in woodwork or a pyrethroid insecticidal lacquer applied, to prevent reinfestation.

Flour Beetles

The name Flour Beetle is usually applied to the small reddish brown beetles of the *Tribolium* species – *T. castaneum*, the aptly named *Red Rust Beetle,* and the oddly named Confused Beetle or *T. confusum*.

Both beetles are common pests of flour and cereal products. Flour that is infested with the beetles smells sour and is unsuitable for bread making as it does not rise satisfactorily.

Flour beetles multiply rapidly, often producing five generations within a year. Their eggs stick to flour particles and hatch out to yellow brown larvae which seek out their food in larders and food stores.

Grain Weevil

Weevils are easily distinguished from other beetles by their long snouts, used for boring into plant food.The Grain Weevil, *Sitophilus granarius,* a brown to black beetle with the characteristic snout, is a major pest in grain stores and flour mills. It sometimes gets transferred to food stores in sacks of flour.

Fig 13.4 Grain weavil *Courtesy: Rentokil*

Pea and Bean Beetle

The Pea and Bean Beetle, *Callosobruchus chinensis,* belongs to a family of seed eating beetles. They have a weevil-like appearance but lack the characteristic snout. The larvae attack the seeds while still on the plant and spend their whole lives within the seeds. They are found in many types of dried pulses such as peas, beans and lentils.

Control

Treat Flour Beetles, Grain Weevils, and Pea and Bean Beetles in the same way:

- Destroy any infested food.
- Clean the store thoroughly.
- Apply an insecticidal lacquer to deter reinfestation.

Silver fish

Silver fish, *Lepisma saccharina*, are primitive, wingless insects which hide away in cracks in warm, damp places in buildings. They are often found trapped in sinks and baths. Their diet consists of carbohydrate materials such as starches, wallpaper paste and cotton materials, supplemented with protein from dead insects and the glue used in book bindings.

The eggs are laid in crevices and hatch into pale coloured nymphs which take 3–4 months to mature. The adult moults throughout life and is exceptionally long lived, surviving for more than three years in optimal conditions. Silver fish are not known to be a health hazard, but their presence suggests that there may be defects in a building, such as leaky pipes or defective ventilation, causing the moist conditions which favour their mode of life.

Control of silver fish

The first step is to check for defective damp courses, condensation problems, or leaky plumbing and to remedy any of these faults. The cracks which give refuge to these creatures can then be treated with carbaryl insect dust to combat the infestation.

Mites

These small pests are a serious nuisance in food stores as they attack a variety of foods such as flour, cereals, cheese and dried fruit. The infested foods have a minty smell and are surrounded with a halo of dust created when the mites were feeding.

Mites belong to the same group as the spiders and, like them, have eight legs when adult. They begin life as eggs laid in crevices and hatch into a six legged larva. The larvae grow and moult several times before they become adult. Some kinds of mite can survive difficult conditions for many months in a cyst-like form called a hypopus.

Control of mites

Still, moist conditions encourage mite infestation so larders and food stores should be well ventilated. If mites are detected, the affected foods should be cleared out and burnt, and the food room *thoroughly* cleaned. Crevices should be treated with a pyrethroid spray and an insecticidal lacquer applied to the walls of the food stores in the same way as for cockroaches.

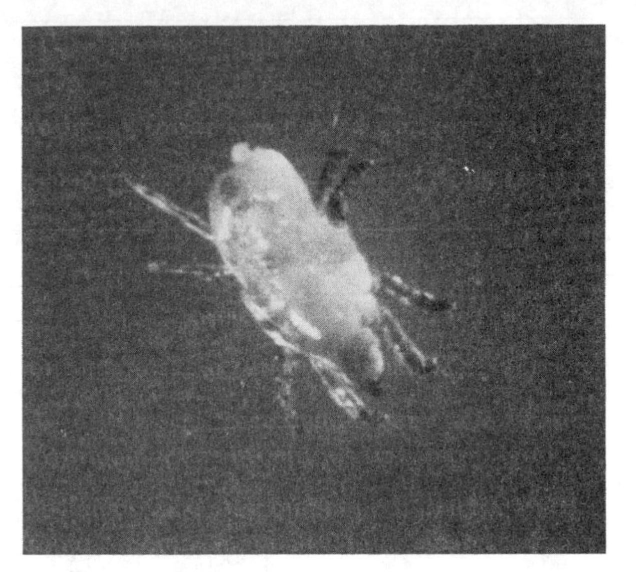

Fig 13.5 Flour mite

Self assessment questions

1 List 4 signs which indicate the presence of rodents in food premises.

2 List 4 measures which can be taken to make food stores rodent-proof.

3 Give 3 reasons why birds should be excluded from food premises.

4 (a) Why are houseflies regarded as the most dangerous insect pests of catering premises?
  (b) Describe measures that can be taken to keep flies out of food rooms.
  (c) How does a insectocutor work? What precautions should be taken in using this type of device?

5 (a) What conditions are likely to attract cockroaches to a food building?
  (b) List the signs that indicate the presence of cockroaches in a kitchen.
  (c) Describe the main ways of controlling cockroaches.

6 (a) How would you know a food was infested by mites?
  (b) How would you deal with a mite infestation in a food store?

7 What sorts of conditions encourage silver fish infestation in a building?

8 (a) What signs would alert you to the presence of Flour Beetles in your food store?
  (b) What measures could you take to get rid of the infestation?

Exercise

Regular pest audits can help to keep infestation to a minimum. Below, is a suggestion for a pro forma which could be used for this purpose. Try it out in your kitchen, store rooms and swill area. You will probably wish to add other questions to suit your particular business.

Pest Control Audit Form

Kitchen

Before starting work in the morning, examine the floor areas around the stoves and machinery.

Are there any signs of food debris? No ☐ Yes ☐ Location? ___

Are there any unclean parts of the stoves? No ☐ Yes ☐ Location? ___

Any parts of machinery unclean? No ☐ Yes ☐ Machine/part ___

Storerooms, larders

Are there any signs of rodent/insect infestation?

Any rodent dropping found? No ☐ Yes ☐ Location ___

Any tracks in dust/flour? No ☐ Yes ☐ Location ___

Any signs of insect damage in – pasta? No ☐ Yes ☐ Details ___

– flour? No ☐ Yes ☐ Details ___

– pulses? No ☐ Yes ☐ Details ___

Any signs of 'nuisance' insects, ants, wasps, fruit flies? No ☐ Yes ☐ Details ___

Swill area

Have a good look round the swill area,

Any bins without lids? No ☐ Yes ☐ Details ___

Any bins overfilled? No ☐ Yes ☐ Details ___

Any signs of rodent or bird infestation? No ☐ Yes ☐ Details ___

Is the floor area clean? No ☐ Yes ☐ Details ___

14 Food poisoning case studies

This chapter consists of case studies which are based on outbreaks of food poisoning which have taken place in the UK or abroad. Each case study consists of three parts:

- A detailed account of the outbreak – the foods involved, their preparation and cooking, and the symptoms of the people affected by the illness.
- Questions on the details of the outbreak. These cover the characteristics of the organism, the type of food poisoning, and the errors in food handling which led to the outbreak.
- You will also be asked to suggest suitable alternative methods of preparation to prevent similar outbreaks.

The questions test your grasp of the facts about the common food poisoning organisms described in Chapter 5 and the correct preparation and cooking methods which are covered in Chapters 6 and 7.

Read through each account carefully and then *write* your answers to the accompanying questions – note form will do. Check your answers in the Discussion sections at the end of the chapter. The answers are highlighted and given a marking value to enable you to assess your understanding and knowledge. The extra information given in the discussions will provide background and amplify the answers.

Case study 1: The tennis tournament

The account

One hot, humid Saturday in June, a suburban club held a tennis tournament. A tea was served at 5 o'clock, consisting of cold roast chicken, salad, cakes, fruit jelly and ice-cream. Some of the remaining portions of chicken were eaten at the supper-dance in the evening.

Early on Sunday, a number of the tennis club members were taken ill with abdominal pains and diarrhoea. Some also had fever, headaches and vomiting.

When the outbreak was investigated, the illness was traced to chicken carcases which were infected at the packing station with *Salmonella virchow*.

The chickens had been stored in a freezer overnight. The owner of the cook-shop removed them at eight o'clock next morning and left them to thaw at room temperature for two hours. The chickens were cooked on rotating spit-roasters for one and a half hours, then left in the kitchen for

30 minutes to cool. The birds were cut by hand into quarters and immediately wrapped – still warm – in grease proof paper. The chicken portions were packed into two cardboard boxes.

These boxes, containing the 120 portions ordered by the club, were delivered at 1.10 p.m. and were kept in the pavillion until 4.30 p.m. when they were unpacked and eaten for tea.

When the kitchen was examined, the food hygiene standard was found to be very poor. The space was inadequate for the amount of work undertaken, so the same work surfaces had been used for both the raw and cooked chickens. In addition, the deep-freeze cabinet was found to contain cooked meats as well as raw chickens.

Questions

1 The Environmental Health Officers suspected *Salmonella* to be the cause of the outbreak even before they completed their detailed investigations. List the details in the account which would have led them to this conclusion.

2 The chickens were infected when the cook-shop owner received them. Nevertheless, the outbreak could have been avoided if the birds had been handled correctly. Make a list of the mistakes made in the preparation, packing and holding of the poultry and suggest more appropriate methods to ensure safe production of the food.

Turn to the discussion on p 193 and assess your answers using the marking scheme.

Case study 2: A hospital outbreak

The account Patients in a Luton hospital were served a 'minced diet' for breakfast at 7.30 a.m. By 6 p.m. some of them were showing symptoms of food poisoning. They all suffered from diarrhoea, but many patients also had pain and vomiting. None were feverish. *Clostridium perfringens* infection was suspected.

Minced ham was thought to be the cause of the outbreak. On the previous day, eleven 5.5kg hams had been boned and rolled and then cooked by steaming. The meat was allowed to 'cool' in a central-heated room and then minced about four hours later. The mincers and choppers were used for both the raw and cooked meats. The mince was refrigerated until 6 a.m. the next day when it was needed for preparing the breakfasts. The mince was mixed with water and soup powder and heated in large steam-jacketed cauldrons.

The food was distributed from the central kitchens to the wards in three vehicles which were electrically heated before leaving the kitchens and cooled gradually en route – for up to an hour. There was some further delay before serving the food from electric hotplates in the wards.

After the outbreak, a kitchen inspection was carried out and the following facts were noted:

- The kitchens were old and not well arranged.
- Raw meat residues were found on the mincers.
- There was an accumulation of unwashed pans.
- The refrigerators were not well ordered. In one there was a cooked leg of beef, much larger than the safe maximum weight of 3kg.
- A room open to the main kitchen and heated by two radiators was used for the pre-refrigeration 'cooling' of cooked dishes.
- The staff toilets in an outbuilding were unclean and one was without toilet paper. Although paper towels were available, a dirty roller towel hung in the corner.

Questions

1 List the aspects of the outbreak that made the investigators make a preliminary guess that the cause was *Clostridium perfringens.*

2 List the errors which were made in the preparation of the meal and give a suitable alternative method in each case:
(a) cooking **(b)** cooling **(c)** mincing **(d)** serving

3 If the reconstituted mince had been brought to a boil, would the *C. perfringens* spores have been destroyed?

The discussion of this study is on p 194.

Case study 3: A cereal story

The account Three people purchased a meal at a Chinese restaurant. They ordered curried shrimps, bean shoots and fried rice. Two of them ate all the food and were taken ill with vomiting and diarrhoea within four hours of finishing their meals. The third person did not like the smell of the rice, so only ate the shrimps and been shoots. He remained well and so did a fourth man who ate chicken and chips. Several other cases of food poisoning occurred after this incident. All were related to food bought at the restaurant on Monday or Thursday.

On investigation, a large number of aerobic sporing organisms were discovered in the fried rice and the boiled rice set aside for frying. Only a few spores, later identified as *Bacillus cereus,* were found in the freshly boiled rice.

The rice for the meals had been boiled, then washed several times and set aside to dry. When needed for frying, it had been placed in the pan with the beaten egg and heated for about a minute. Often the rice was left to dry overnight and used for frying on the following day. Any boiled rice left over was added to the next batch.

The restaurant closed on Wednesdays and was open for four hours on Sundays.

Questions

1 Rice is usually classified as a 'safe food'. Explain why it caused this outbreak of food poisoning.

2 Give details of a safe method of cooking rice for a dish of the kind described.

3 In what ways are *Bacillus cereus* and *Clostridium perfringens*:
- **(a)** similar to each other?
- **(b)** different from each other?

You will find the discussion for this study on p 196.

Case study 4: The school dinner

The account A cold chicken dinner was served in a school one day during the summer term. Less than three hours after the meal, 133 of the 276 diners suffered severe vomiting and some had diarrhoea.

The chickens had been cooked on the previous day. The meat had been removed by hand from the bone, sliced and refrigerated overnight.

When the epidemic was investigated, the same strain of *Staphylococcus aureus* was found in samples taken from the nose of the cook, the hands of the assistant cook, the lining of the refrigerator, and in specimens from several of the people who were ill. The temperature range of the refrigerator where the chicken had been stored was found to be 10°– 15°C.

Questions

1 Like many healthy people, the two cooks carried *Staphylococcus aureus* in their noses and on their hands. List the precautions they should have taken to prevent these bacteria from being transferred to the food they were preparing.

2 What instructions do you suggest should be given to the kitchen staff as regards the care of the refrigerator in the future?

The discussion for this study is on p 197.

Case study 5. The fatal fish supper

The account Three people shared a supper which included canned sprats. Before the meal, they had debated whether to eat the fish, as the can had shown a definite bulge and gas had escaped when it was opened. The fish had not smelt or tasted bad and one man had eaten twelve fish, another six, and the woman who shared the meal had eaten three.

The following day, the man who had eaten the largest quantity of fish became ill suddenly. He felt dizzy, breathed with difficulty and began to see double. Soon his muscles became paralysed, and the next day he died. Later, the second man also developed symptoms of botulism and so did the woman, though she was less acutely ill. They were both very lucky to recover from the disease as the antitoxins A and B given to them proved ineffective in neutralising the toxin in the fish.

Questions

1 List the external signs which suggest the contents of a can of fish, meat or vegetables might be dangerous. What should you do if you detect such signs in a can you are about to open?

2 What precautions should you take when preparing root vegetables in the kitchen?

3 Why is it dangerous to can or bottle fish, meat or vegetables at home?

The discussion for this study is on p 198.

Discussion of case study 1

1 *Salmonella* poisoning was suspected because:

- **The incubation period was 10–12 hours.**
- **The symptoms were diarrhoea and abdominal pain and in some cases fever.**
- **The suspected food was chicken.** **Marks 4**

None of these facts on their own would prove the outbreak was due to *Salmonella*. To do this with certainty, the bacteria have to be isolated from the suspected food and from samples taken from the affected people. Then the bacteria have be cultured in the laboratory and tested to establish their identity. This takes time, so it is usual to make a preliminary diagnosis based on the length of the incubation time, symptoms of the disease, and the type of foods suspected of causing the outbreak.

2 *Salmonellae* are very common organisms in poultry. Since the birds are raised in hatcheries containing thousands of birds the opportunity for many birds to be infected is always present.

Points to be noted from the account are:

- **The defrosting time was too short.**

Marks 2

The investigators found 8–10 hours were necessary to effect complete thawing of the chickens at room temperature.

- **The cooking method would not have killed *Salmonellae* inside the chicken.**

Marks 2

The inner parts of the chickens would have been frozen when they were placed on the spit. The time and the temperature reached by this method of cooking would be inadequate to be sure of killing all the *Salmonell*a bacteria.

Oven roasting after complete thawing would have been satisfactory (190°– 204°C, 30 minutes per 450g weight)

- **The cooked birds were cut up on the same surface as the raw birds.**

Marks 2

This provided an opportunity for the cooked birds to be re-infected. In a well-ordered kitchen, separate areas would be set aside for handling raw and cooked foods, and surfaces and equipment would be thoroughly cleaned after use.

- **The chicken joints were wrapped in grease-proof paper whilst still warm.**

Marks 2

This would have produced warm, moist conditions around the meat, ideal for the growth of organisms. They should have been cooled quickly and stored in a refrigerator until needed and wrapped when cold.

- **The chicken was delivered far too early for a meal at 5 o'clock.**

Marks 1

The chicken had over four hours to stand in hot, humid conditions. If refrigeration was not available at the club, the food should have been delivered just prior to the preparation of the meal.

- **The chicken should not have been held over for a second meal.**

Marks 1

This provided a further period of time for multiplication of organisms. A second batch of food should have been supplied later in the day.

- **Cooked chicken should not have been stored in the same freezer as the raw chicken carcases.** **Marks 1**

Again there was an opportunity for cross infection.

Did you spot most of the mistakes? **Total marks 15**

Discussion of case study 2

1 *Clostridium perfringens* was suspected because:

- **The incubation period was about $10\frac{1}{2}$ hours.**

- **All the patients suffered from diarrhoea, some had pain and vomiting but none were feverish.**
- **The meal had been prepared the previous day and reheated.**

Marks 3

The fact that the patients were not feverish suggested *C. perfringens* rather than *Salmonella* and the incubation period was nearer to that of *C. perfringens*. The fact that a large quantity of a made-up meat dish had been reheated would also tend to suggest *C. perfringens* as the probable cause.

2(a) Cooking

- **The ham joints were too large.** **Marks 1**

Cooking such large joints was hazardous particularly as the joints were boned and rolled. This process carries infection from the *outside* of the meat to the *centre* where it easily escapes destruction.

- **The hams were steamed.** **Marks 1**

Steaming does not allow enough heat penetration to kill spores in the centre of large joints.

- **Smaller joints, cooked by roasting, would have ensured that the *C. perfringens* spores were killed.** **Marks 2**

Note: The hams were specifically intended to make **minced diets,** so **the ham could have been minced raw and then cooked and served hot.** This would have cut out all the hazardous stages of preparation.

(Give yourself 5 bonus points if you spotted this important point.)

(b) Cooling

- **The hams were left to 'cool' in a centrally heated room so their temperatures would have dropped very slowly.** **Marks 1**

As soon as the hams had dropped below 50°C, the spores surviving in the meat would have been able to germinate and multiply.

- **The hams, if cooked as joints should have been placed in a well ventilated cold room to cool as rapidly as possible and then stored in a refrigerator.** **Marks 2**

(c) Mincing

- **Mincing the ham would have spread the infection from the centre of the joints throughout the meat.** **Marks 1**
- **Raw and cooked meats should be prepared with different mincers and choppers, or the equipment washed and sterilised between uses.** **Marks 4**

All tools, machines and food preparation surfaces need thorough cleaning and disinfection after use.

(d) Serving

- **The service was too slow and allowed the food to cool to a dangerous degree.** **Marks 1**
- **The food should have been served promptly at a temperature above 63°C.** **Marks 2**

Safe service of food in a large establishment like a hospital is always a problem. The difficulties can be lessened by good liaison between kitchen and service staff so that food is sent out hot and served quickly.

Electrically heated trolleys need frequent checks to ensure that they maintain the food at a safe temperature. This is particularly important in a hospital since babies, sick people and the elderly are the most vulnerable to food poisoning. Illness which would cause only temporary distress in normal healthy people can be fatal to these sections of the population.

3

- **Heat resistant *C. perfringens* spores survive more than 30 minutes of boiling and many survive much longer.** **Marks 2**

In cooked meats, where spores are protected by proteins, some strains may survive up to 5 hours at 100°C. We can only be sure of killing *C. perfringens* spores in joints (3 kg or under) if we use methods such as roasting or pressure cooking which raise the temperature well above the boiling point of water.

Total marks 20 + 5 bonus points

Discussion of case study 3

1

- **Rice is only classified as a 'safe' food when it is in the dry condition. In this state it does not contain sufficient water to allow growth of organisms.**

Marks 2

The dry rice in this case contained small numbers of *B. cereus* spores. They were not able to multiply in the dry rice nor were they present in any larger numbers in the *freshly* boiled rice.

- **Since the rice had absorbed a large amount of water during cooking, this allowed the spores to germinate and multiply when it was kept overnight at room temperature.**

Marks 2

2

- **Boil the rice in batches as it is needed for frying. In this way, it is unlikely that there will be any boiled rice left over at the end of**

the day – if any does remain, it should be thrown away.

Marks 4

3 (a)

- **Both are rod shaped bacteria which produce spores.**

Marks 1

(b)

- ***C. Perfringens* is an anaerobe i.e. its growth is inhibited by the presence of oxygen.**
- ***Bacillus cereus* is aerobic as a rule but can also grow in the absence of oxygen.**

Marks 1

Total marks 10

This study illustrates that it is just as dangerous to hold cereal foods (rice, custards, cornflour sauces) at warm temperatures as it is to expose meat dishes to these conditions.

Discussion of case study 4

1

- **The chicken meat should have been removed from the bones using clean implements. Preferably, the cooks should have worn disposable plastic gloves.**

Marks 2

- **The meat should have been cooled quickly, covered and placed in the refrigerator.**

Marks 1

The safest course would have been to cook, cool and serve the chicken on the same day. As a fairly large number of people had to be served, this may have been impractical. If the meat had to be prepared in advance, it was essential to prevent transfer of large numbers of organisms and to ensure that any bacteria which did get into the meat could not multiply.

2

- **The refrigerator should have been regularly examined, defrosted and cleaned.**

Marks 2

The presence of *Staphylococci* in the refrigerator lining suggests that it was dirty and neglected. Regular defrosting is necessary to ensure that the refrigerator is working efficiently. Refrigerated foods should be examined regularly and any past their best or beyond their 'eat by date', removed.

The refrigerator should be cleaned out thoroughly using bicarbonate of soda and a mild detergent then dried before the food is replaced.

- **Raw and cooked foods must be kept apart – either in separate refrigerators or, if stored in the same refrigerator, the raw meat must be placed *below* the cooked food. Both must be covered.**

Marks 2

This precaution is essential to prevent cross contamination.

- **Cooked foods must be cooled to 10°C or below before being placed in the refrigerator.** **Mark 1**

Placing hot food in a refrigerator makes the machinery overwork, may raise the temperature above the safe limit and causes condensation on the food as it cools.

- **The temperature of the refrigerator should be checked regularly and adjusted if necessary, to keep it within the correct range of 1°–4°C.** **Marks 2**

Thermostats are fairly crude devices and require periodic adjustment to maintain a steady temperature when the outside temperature changes and the amounts of food inside the refrigerator vary.

Total marks 10

(If you missed some of the points on the refrigerator, turn to p 156 and refresh your memory.)

Discussion of case study 5

1 Cans should be discarded UNOPENED if they show any of the following faults:

- **signs of being 'blown' i.e. one or both of the ends of the can bowing out due to pressure of gas inside**
- **damage at the seams**
- **any serious denting**
- **any serious rusting.** **Marks 4**

These signs show either that the can has been insufficiently processed or has been damaged subsequently so that air or water contaminated the contents.

2

- **The soil should be washed off carefully and the washing done well away from other foods.** **Marks 2**

The soil is the home of a number of food poisoning organisms including *C. botulinum, C. perfringens* and *B. cereus* and their spores can contaminate other foods.

3

- **These are all neutral or low acid foods. They may contain *C. botulinum* spores. A high temperature is needed (121°C) to kill these spores. It is usually not possible to obtain these conditions at home. Marks 4**

C. botulinum does not grow in acid conditions (pH 4.5 or below) so it is quite safe to bottle or can acid fruits at home.

Total marks 10

(If you have forgotten about the details of the pH scale turn to p 49. If you are not sure about the pH values of different types of food consult the table on p 50.)

References

The facts for the case studies were taken from the following sources:

Case study 1
A.B. Semple *et al. Brit. Med. J.* 1968, 4,801–1803.

Case study 2
Mair Thomas *et al. The Lancet,* May 14, 1977.

Case study 3
P.R. Mortimer, G. McCann. *The Lancet,* May 25, 1974.

Case study 4
Report, Epidemiological Research Laboratory, Public Health Laboratory. *BMJ.*, 7th Aug., 1974.

Case study 5
Elizabeth L. Hazen. *J. Infect, Diseases,* 1937, 260–264, 60.

Index